Foundations for Undergraduate Research in Mathematics

Series Editor

Aaron Wootton, Department of Mathematics, University of Portland, Portland, USA

Eli E. Goldwyn • Sandy Ganzell •
Aaron Wootton

Editors

Mathematics Research for the Beginning Student, Volume 1

Accessible Projects for Students Before Calculus

 Birkhäuser

Editors
Eli E. Goldwyn
Mathematics
University of Portland
Portland, OR, USA

Sandy Ganzell
Mathematics & Computer Science
St. Mary's College of Maryland
St. Mary's City, MD, USA

Aaron Wootton
Mathematics
University of Portland
Portland, OR, USA

ISSN 2520-1212   ISSN 2520-1220   (electronic)
Foundations for Undergraduate Research in Mathematics
ISBN 978-3-031-08559-8   ISBN 978-3-031-08560-4   (eBook)
https://doi.org/10.1007/978-3-031-08560-4

Mathematics Subject Classification: 00A08, 00B10, 00-02

This book is published under the imprint Birkhäuser, www.birkhauser-science.com by the registered company Springer Nature Switzerland AG
The registered company address is: Gewerbestrasse 11, 6330 Cham, Switzerland

Preface

It is our strong belief that opportunities for mathematical research should be available to all students with an interest in mathematics—not just to those who already have expertise in the field. In our experience, engaging with research has led students to discover a talent and passion for mathematics they didn't even know they had. In fact, studies show that students who participate in early-career STEM research are more likely to remain in school, are more likely to remain in a STEM major, perform better in upper division courses, and are more interested in postgraduate STEM educational opportunities.[1] Research experience is also a desirable trait in private industry, as it illustrates an individual's abilities to problem solve.

Though there has been significant growth in research experiences in mathematics for undergraduates at 4-year colleges and research universities, these opportunities are often only available to upper-level students or those with significant background knowledge. Similar research experiences in mathematics for the beginning student—that is, those in community college and early career college students—are much rarer, even though this group stands to benefit significantly from such opportunities.

Perhaps the biggest barrier faced by the beginning student and their faculty mentors is not knowing where to begin. Accessible topics for new student researchers are often hard to find, even for experienced teachers, and finding unanswered questions that are well suited to student projects is time consuming. Accordingly, the main goal of these two volumes is to expand research opportunities in mathematics for the beginning student by removing this significant hurdle. Specifically, we seek to provide community college students, early college students, and perhaps even advanced high school students everything they need to initiate research projects with or without a faculty mentor, and to foster independence in research.

Each chapter in these two volumes is a self-contained, accessible article that provides ample background material, recommendations for further reading, and perhaps most importantly, specific projects that can be pursued immediately upon reading the chapter. What makes these volumes different from other FURM volumes

---

[1] Hewlett, James A. *Broadening participation in undergraduate research experiences (UREs): The expanding role of the community college.* CBE-Life Sciences Education 17.3 (2018): es9.

is that the chapters have been written for the beginning student by minimizing the number of prerequisites. Indeed, many of the chapters require no prerequisites other than a desire to pursue mathematics, and even the most ambitious chapters require no more than linear algebra or introduction to proofs, both of which are typically sophomore-level classes.

This volume is geared towards students with minimal experience beyond traditional high school mathematics. For the reader's ease, we roughly order the chapters by prerequisite. The first four chapters require only high school algebra, the fifth chapter precalculus, the sixth a little probability and programming, the seventh and eighth some exposure to calculus, and the ninth a brief introduction to proof.

In Chap. 1, "Games on Graphs," Carlson, Harris, Hollander, and Insko introduce three games on graphs: cops and robbers, hungry spiders, and broadcast domination. Each of these games allows for a multitude of research opportunities and further generalizations; the games can be studied and analyzed from a variety of perspectives, providing students an immediate area of research with very little need for much technical or sophisticated mathematical background. Anyone who enjoys playing board or video games should read this chapter and think about how their favorite games can be modeled as a game on a graph.

In Chap. 2, "Mathematics for Sustainable Humanity—Population, Climate, Energy, Economy, Policy, and Social Justice," Wang takes the readers to confront global challenges facing humanity and our planet. Through a sequence of sustainability and mathematical concepts, exercises, and real-world projects, the reader is brought along on this journey to develop an understanding of the complex environmental, economic, and sociocultural interlinkages; to create powerful data visualizations; to interpret and draw inferences from functions, equations, models, and graphs; and to use the knowledge to shape competency and reach justified decisions.

In Chap. 3, "Mosaics and Virtual Knots," Ganzell introduces the mathematics of knots. Of course, knots are important to sailors and rock climbers, but interesting properties of knots can also be discovered from simple diagrams called mosaics, formed by placing square tiles onto a grid. No mathematical knowledge other than basic arithmetic is needed. After learning some of the notation for drawing and describing knots, the reader will explore how to create these mosaics, trying to be as efficient as possible. Then Ganzell introduces an abstraction known as a virtual knot. Using mosaics to describe virtual knots is a relatively new idea with lots of open questions to pursue.

In Chap. 4, "Graph Labelings: A Prime Area to Explore," Donovan and Wiglesworth introduce the reader to prime labelings of graphs. Research in prime labelings asks the question, "Can the vertices of a graph be labeled in such a way that the labels of adjacent vertices are relatively prime?" Several variations of prime labelings are then introduced while the authors guide the reader through examples and existing research. A variety of research questions for each modification, a few of which have never been explored, are presented, thus allowing the reader to make progress on new problems.

In Chap. 5, "Acrobatics in a Parametric Arena," Shelton, Henderson, and Gebhardt introduce parametric functions in an engaging and accessible way, examining component graphs with single variable calculus, and indicating online resources to complete some exercises. To illustrate, they analyze sinusoidal fits to original 2D data from motion capture of actual acrobatic movements of juggling sticks. They also analyze public data of cell phone subscriptions to juggle shares in a competitive financial market, exploring various fits to the 3D data. A number of research projects are proposed to inspire the reader to perform their own imaginative parametric acrobatics.

In Chap. 6, "But Who *Should* Have Won? Simulating Outcomes of Judging Protocols and Ranking Systems," Lewis and Clifton introduce the mathematical and computational skills needed to address the issues of fairness and bias in various contests. A brief introduction to probability is provided, followed by instruction for performing simulation experiments in R (other programming software can easily be substituted). Detailed examples and code snippets provide the reader with the skills needed to tackle any of the five suggested research projects, ranging from discovering how likely it is that the two highest-ranked teams going into a tournament end up in the championship playoff to investigating the stability and fairness of three different voting schemes for an election. Each of the proposed projects can be adapted to suit the interests of the reader.

In Chap. 7, "Modeling of Biological Systems: From Algebra to Calculus and Computer Simulations," Dimitrov et al. present an expansive view of mathematical tools as used in modeling a variety of biological and biomedical systems. The first two projects model a biological and a biomedical system and are accessible to anybody with knowledge of high school algebra. The final two projects build on more advanced population and infectious disease models and require some knowledge of calculus. The chapter introduces computational thinking and computer-based simulations as important tools for approaching problems in modern applied mathematics. Mathematical modelers leverage mathematical techniques with those computational tools to tackle problems that are much more complex than previously possible, allowing them to address more realistic and relevant societal issues with their modeling efforts.

With the massive impact that COVID-19 has had on the lives of students, mathematical epidemiology is a particularly timely subject. In Chap. 8, "Population Dynamics of Infectious Diseases," Ledder and Homp present an individual-based epidemic model and a continuous-time SEIR epidemic model, using a presentation that is accessible to students who have a strong background in algebra and functions, with or without calculus. Projects focus primarily on extending the models to novel situations, including more complicated disease histories and incorporation of mitigating strategies such as isolation and vaccination. The projects are categorized according to the levels of calculus and programming that are required.

In Chap. 9, "Playing with Knots," readers will learn about how mathematicians think about knots. Given two knots, how can you tell if they're the same or different? How hard is it to unknot certain knots, and how can we measure a knot's complexity? As if that weren't intriguing enough, Henrich describes several games that can be played with knots. She reveals the winning strategies researchers have uncovered to ensure that, when playing knot games against unwitting competitors, they can always win. But researchers have only just scratched the surface of learning about knot games. Can *you* discover winning strategies for games on knots that have not yet been explored?

Portland, OR, USA   Eli E. Goldwyn
St. Mary's City, MD, USA   Sandy Ganzell
Portland, OR, USA   Aaron Wootton

Contents

Games on Graphs: Cop and Robber, Hungry Spiders, and Broadcast Domination

Joshua Carlson, Pamela E. Harris, Peter Hollander, and Erik Insko

Abstract

There are many types of games one can play on a graph. In this chapter, we introduce three games on graphs: cops and robbers, hungry spiders, and broadcast domination. Each of these games allows for a multitude of research opportunities and further generalizations of the game, and the games can be studied and analyzed from a variety of perspectives, providing for students an immediate area of research with very little need for much technical or sophisticated mathematical background. We end the chapter by describing some courses beyond calculus and additional techniques they provide, which will allow for further study of these games from new mathematical perspectives.

Suggested Prerequisites *Curiosity and a love for games. A very basic background on graphs.*

J. Carlson (✉)
Drake University, Des Moines, IA, USA
e-mail: joshua.carlson@drake.edu

P. E. Harris
University of Wisconsin Milwaukee, Milwaukee, WI, USA
e-mail: peharris@uwm.edu

P. Hollander
Williams College, Williamstown, MA, USA
e-mail: pjh1@williams.edu

E. Insko
Florida Gulf Coast University, Fort Myers, FL, USA
e-mail: einsko@fgcu.edu

© The Author(s), under exclusive license to Springer Nature Switzerland AG 2022
E. E. Goldwyn et al. (eds.), *Mathematics Research for the Beginning Student,*
Volume 1, Foundations for Undergraduate Research in Mathematics,
https://doi.org/10.1007/978-3-031-08560-4_1

1   Introduction

Since ancient times, humans have been fascinated with games. In fact, games form an integral part of every human culture, and they provide us with a way to both exercise our intellect and socialize with others. It is no surprise, then, that mathematicians have studied ways to identify winning strategies, measure game complexity, or determine the unsolvability of games and puzzles. Luckily for us, there is absolutely no shortage of games in which such analyses can be performed. Also, as most games are often simple enough to introduce, yet challenging enough to analyze and/or solve mathematically, this setup allows a wide breadth of people with varying mathematical backgrounds to enjoy both the play and the analysis.

In this chapter, we consider games of two types: two-player and single-player games. In two-player games, two players alternate turns playing the game, and they each want to figure out a way to win the game. Classical examples of two-player games include tic-tac-toe, checkers, and chess. As expected, a single-player game is one in which a single player is given some information or resources, and they play a game to see if they can win by solving a task. The card game solitaire is an example of a single-player game. Moreover, games can be played using cards, tokens, or boards with pieces.

In our work, we focus on games we can play on a mathematical structure known as a graph, which is made up of vertices (dots) and edges (line segments) connecting the vertices. In order to make our approach concise, we delay all of the technical definitions and background related to graphs until Sect. 6, and any further definitions needed for a particular game are presented in their corresponding section. The three games we present include a two-player game, a single-player game, and a game that can be either.

Our first game is known as the *cop and robber game*. In this game, we place a "cop" and a "robber" on the vertices of a graph. The game begins with the cop moving from their starting vertex to an adjacent vertex (one connected via an edge), which is followed by the robber moving to an adjacent vertex from where they began. The goal of the game is to determine whether the cop can catch the robber, which is when the game ends. The second game we present is called *hungry spiders*, where one or more flies are stuck at fixed vertices in a web (a finite graph), and one or more spiders try to capture them.[1] Each round, the player who is trying to save the flies, gets to choose one edge of the graph to delete in each round, and the game ends when the player deletes enough edges to separate the spiders from the flies or when the spiders capture a fly. The last game we present is a single-player game called *broadcast domination* in which a single-player allocates "resources" on certain vertices (called "towers"). These "resources" are then propagated to neighboring vertices, and the ultimate goal of the game is to find an efficient placement of as few "towers" as possible while spreading sufficient resources to every vertex on the graph.

---

[1] Our hungry spiders game is based on the Greedy Spiders mobile app game.

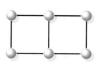

Fig. 1 Cycle graph C_6, wheel graph W_6, and grid graph $G_{2,3}$

Now that we know some initial information about our games, let our game playing begin!

Game 1: Cops and Robbers Select a graph from Fig. 1, draw it large on a piece of paper and grab a friend. Find two different tokens (coins or beads work great for this).

Now, here are the rules: Player 1 places their token down on a vertex, and then Player 2 places their token down on some other vertex. The players then alternate moving their tokens along the edges to a vertex next to the vertex they are currently on. Player 1 wins the game if there is ever a time when the tokens are on the same vertex, and Player 2 wins the game if this can always be avoided. Who wins the game? What if you try a different graph?

Game 2: Hungry Spiders Select a graph from Fig. 1, draw it large on a piece of paper, grab a friend, and two or three different tokens. Now, here are the rules: Player 1 (the spider) places a token down on a vertex, and then Player 2 (the fly) places a token down on some other vertex. The players then alternate turns, starting with the fly. In each round of the game, the fly gets to remove one edge from the graph, and the spider gets to move from the vertex they currently occupy along an edge to an adjacent vertex. The spider wins the game if there is ever a time when the tokens are on the same vertex, and the fly wins if there are no longer any paths that the spider can follow to reach the fly. Who wins the game? What if you try a different graph? Does the outcome depend on where the players start? What if Player 2 has to place two flies on the graph at the start? What if Player 1 gets to place two spiders? As you can see, there are many possible generalizations to make the game interesting, and we believe that studying these variations could provide a wealth of undergraduate research opportunities.

We remark that we provide Sect. 6 with a short primer on graph theory background. The reader can, as needed, refer to that section for more technical background, and once the reader begins actively researching a topic, we recommend reading the articles cited in this chapter (and conducting a literature review of the articles they cite as well) to see examples of standard proof techniques in analyzing games on graphs. We are now ready to formally introduce our games on graphs

2   The Game of Cops and Robbers

The original game of cops and robbers on graphs was introduced independently by Quilliot [9] and Nowakowski and Winkler [8]. The game of cops and robbers on a simple connected graph G proceeds as follows. Begin the game at time 0 by placing a cop and a robber on distinct vertices of G. At time 1, the robber moves from its occupied vertex to a new vertex by walking across an edge of the graph G. At time 2, the cop moves in a similar fashion. As the cop and robber alternate turns moving from vertex to adjacent vertex in $V(G)$, the cop is trying to capture the robber, and the robber is trying to elude the cop. The game comes to an end when the cop and the robber occupy the same vertex, in which case the robber is caught. An example of a sequence of moves (or turns) for a game of cops and robbers is illustrated in Fig. 2.

Challenge Problem 1 Consider the game of cops and robbers on the path graph with n vertices. If the robber and cop are placed on this graph, and the smallest number of vertices between them is m (i.e., the distance between their respective positions is $m + 1$), then what is the maximum time the game will last when the cop is playing optimally (always trying to reduce the distance between themselves and the robber)?

The previous challenge problem may lead us to incorrectly assume that the cop can always catch the robber. However, even when playing optimally this might not be the case! Consider playing the game on the cycle graph with $n \geq 4$ vertices. In this case, so long as the starting distance between the cop and the robber is greater than one, the cop will chase the robber around the graph forever and the game will never end. We call such a graph a "robber win" graph.

> **Research Project 1** Consider graphs on at most 10 vertices (a list of many interesting graphs is found here: https://www.graphclasses.org/smallgraphs.html), and determine which are robber win graphs.

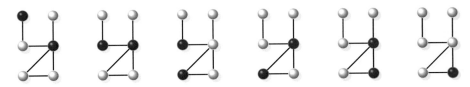

Fig. 2 A game of cops and robbers that ends at time 5 (cop is blue and robber is red)

Challenge Problem 2 Consider a variation of the game of cops and robbers in which the robber can choose to move or to remain on the same vertex each turn, while the cop always moves on their turn trying to catch the robber. Using this version of the game, consider again Challenge Problem 1, and determine the maximum time the game will last under this new constraint.

In the remainder of the section, we consider a change to the way in which the cop and robber can move as well as the game's objective. In this variant, the initial placements of the cop and robber occur as follows. First, the cop chooses an initial vertex to occupy. Then, the robber can observe the initial placement of the cop and use this information as they choose their initial vertex. For movement, suppose the cop and the robber both are able to either remain on the same vertex or freely move to any neighboring vertex on their respective turns. As for the objective, first, consider the following notion of damage introduced by Cox and Sanaei in 2019 [3].

A vertex v is said to be *damaged* whenever the robber is able to make a move from v (either staying on v or moving to a neighbor of v) without being caught during the cop's next turn. Once vertices are damaged, they cannot be undamaged. Also note that from the robber's perspective, simply moving to a vertex v is not sufficient for damaging that vertex. The robber must be able to take a turn while starting on v without immediately getting caught by the cop. For example, suppose the cop and robber start at opposite ends of the path P_{10} as in Fig. 3. If the players always move toward each other on each turn, only four vertices are damaged since the robber is caught before they can make a move from v_6.

The idea of the *damage variant of the cops and robbers game* is that, aside from capturing the robber, minimizing damage might be the next best thing that the cop can do. In this variant, optimal play for the cop minimizes the number of damaged vertices, and optimal play for the robber maximizes that number. Note that the initial placement step plays a role in considering optimal play since the wrong initial placement could influence the number of damaged vertices. In [3], the authors define the *damage number* of a graph G, denoted $dmg(G)$, as the minimum number of vertices damaged in G over all games played where the robber plays optimally. In other words, $dmg(G)$ is the exact number of vertices damaged in G when both the cop and the robber play optimally. As a consequence, determining the damage number does not always require capture of the robber.

Observe that for a given graph G and integer k, we can prove that $dmg(G) = k$ in two steps. First, we can prove $dmg(G) \le k$ by producing a strategy for the cop that guarantees that no more than k vertices can be damaged regardless of the robber's strategy. Second, we can prove $dmg(G) \ge k$ by producing a strategy for the robber that guarantees that at least k vertices are damaged regardless of the cop's strategy.

v_6

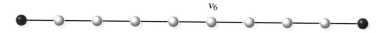

Fig. 3 The path P_{10} with cop (blue) and robber (red) starting at opposite endpoints

Fig. 4 A graph G with
$\mathrm{dmg}(G) = 1$

Challenge Problem 3 For each integer $n \geq 1$, determine with proof $\mathrm{dmg}(P_n)$.

Challenge Problem 4 For each integer $n \geq 4$, describe a strategy for the cop that
proves $\mathrm{dmg}(C_n) \leq \lfloor \frac{n-1}{2} \rfloor$.

An interesting property of the damage variant of the cop and robber game is that
even though the cop may be able to catch the robber, this may not be optimal for
minimizing damage. For example, in the graph in Fig. 4, there is a strategy for the
cop that guarantees capture of the robber in which multiple vertices are damaged (try
to verify this). However, note that if the cop is initially placed on the blue vertex,
they can catch the robber whenever the robber moves to a white vertex. Therefore, if
the robber starts on a white vertex, no damage is done. Thus, the robber can choose
to start on a red vertex and remain there to achieve maximum damage. Either way,
the cop minimizes damage by staying in place rather than capturing the robber.

Note that by choosing a vertex of maximum degree and remaining on that vertex,
the cop can always protect $\Delta(G) + 1$ vertices. This implies that for any graph G on
n vertices, $\mathrm{dmg}(G) \leq |V(G)| - \Delta(G) - 1$ [3]. We leave it to the reader to verify
the following:

Challenge Problem 5 Graphs on n vertices with $\Delta(G) = n - 1$ or $\Delta(G) = n - 2$
satisfy $\mathrm{dmg}(G) = n - \Delta(G) - 1$.

In [2], the authors show that these are not the only graphs with damage number equal
to $|V(G)| - \Delta(G) - 1$. This motivates the following research project (originally
posed in [3]).

Research Project 2 Give a complete characterization of all graphs G on n
vertices that satisfy $\mathrm{dmg}(G) = n - \Delta(G) - 1$.

We can also play the damage variant with k cops versus a single robber. In this
version, the objectives are the same, and the two teams (cop team and robber team)
alternate taking turns with all k cops moving simultaneously on their turn. The k-
damage number of a graph G, $\mathrm{dmg}_k(G)$, is defined similarly to $\mathrm{dmg}(G)$ except k

cops play against one robber. In [2], Carlson, Eagleton, Geneson, Petrucci, Reinhart, and Sen define the k-damage number, but they do not determine this number exactly for many families of graphs. This leads to the following project.

> **Research Project 3** Suppose we are given an arbitrary positive integer k. For which graphs G, can we determine an exact formula for $dmg_k(G)$ in terms of k and $|V(G)|$?

3   Hungry Spiders

The next set of projects concerns questions motivated by the Greedy Spiders game created in 2011 by the indie game studio Blyts. Since we will be considering variations, we call it the hungry spiders game. In this game, the graph G represents a spider web where a set of flies are stuck at a subset of vertices $F \subseteq V(G)$. The spider(s) are allowed to move about the web following a rule or strategy, traversing one edge at a time during their turn. In Blyts' Greedy Spiders game, the spiders are controlled by the computer, and each spider moves toward a nearest fly in each round of the game. The player gets to remove one edge per turn in an effort to separate the flies from the spider(s). The spider wins if it manages to reach any one of the flies, and the flies win if the spider can no longer reach any of them.

Figure 5 shows a possible web and a sequence of moves in which the fly wins. An interested reader may also watch the game in action in the official trailer using the following link: https://youtu.be/qv0X-m1pbCw.

We now describe the game more formally in the language of graph theory. At the start of the game, the player is presented with a simple graph $G_0 = (V, E_0)$. The spiders are placed at some subset of vertices $S_0 \subseteq V$, and the flies are stuck at a subset $F \subseteq V$ of vertices for the duration of the game. In the original version of the game, the spiders' moves follow a greedy rule, meaning they always move to a vertex that decreases the distance between them and at least one of the nearest flies. (As an aside, this particular greedy rule may not always be the optimal strategy for the spiders to employ, and later in this section we ask what other strategies the spiders might implement in different starting configurations.)

An instance of the game proceeds in rounds, and at the beginning of round i, we call the current graph $G_i = (V, E_i)$ and the current location of the spiders S_i. Each round consists of the following two steps:

Fig. 5 A sequence of moves in which the fly wins (fly in blue and spider in red)

1. The player picks an edge $e \in E_i$ to delete from $G_i = (V, E_i)$, forming the new graph $G_{i+1} = (V, E_{i+1})$
2. The spiders jointly compute their next move according to the greedy rule, and each spider moves to an adjacent vertex. Hence, their set of locations S_i becomes S_{i+1}.

Throughout the game, the player's goal is to find a sequence of edge deletions which keep S_t from F for all $t \geq 0$. In other words, the player wants to find a sequence of edge deletions that disconnects the part of the graph containing the spiders from the part of the graph containing the flies. If any spider ever reaches a fly, the game on that graph is over, and the flies have to try again. Figure 6 shows one such case.

While creating new levels, the game developers want to categorize the difficulty level of various graphs and starting configurations, beginning with the ones that are too easy. For instance, Blyts' game starts with two warm-up levels containing a single fly trapped on a web with a single spider.

Challenge Problem 6 Show that a single fly can win against a single spider on any graph regardless of where the spider is placed.

We note that slight variations in a graph or the placement of the spider and flies on the same graph can make the game much more challenging or change the game's outcome.

Challenge Problem 7 Determine which of the graphs in Fig. 7 are fly-win or spider-win if both the spider and flies employ their optimal strategy.

After playing a few rounds, the reader will surely notice that if the spider can reach a vertex that is adjacent to two or more flies, then the spider will win.

Challenge Problem 8 Construct a family of graphs and configurations of flies on which the flies win when the spider employs a greedy rule that simply moves toward

Fig. 6 A sequence of moves on a graph in which spider wins (flies in blue and spider in red)

Fig. 7 Classify each game board as fly-win or spider-win given these starting configurations

a nearest fly, but not if the spider employs a different strategy that moves toward the nearest vertex that is adjacent to two or more flies. (One such example appears in Fig. 7.)

Research Project 4 Is there a succinct classification of all graphs and configurations of flies on which the flies win when the spider employs a greedy rule that simply moves toward a nearest fly, but not if the spider employs a different strategy that moves toward the nearest vertex that is adjacent to two or more flies?

Finding examples of graphs with a certain property is often easy, but classifying all graphs with that property usually takes a considerable amount of work.

Research Project 5 Classify families of fly configurations on graphs where a set of flies win if they play optimally against a single spider.

We have already noted that a graph with only one fly is always a fly-win graph. One sufficient condition in your classification might be configurations in which the spider cannot reach any vertex of the graph that is adjacent to two or more flies. Another sufficient condition might be configurations in which the flies can cut all of the edges between their component and the spider's before the spider can reach that cut.

Research Project 6 Classify families of player configurations on graphs where a set of flies win if they play optimally against two spiders.

For later levels, the game developers want graphs where the flies win, but only if they make a series of carefully chosen edge deletions.

Challenge Problem 9 Construct a family of graphs on which the flies can win but not without deleting at least four edges. Construct a family of graphs on which the fly can win but not without deleting at least n edges.

Additionally, these types of questions can be explored regarding an even broader set of graphs.

Research Project 7 Tackle the previously stated challenge problems and research projects on *multi-graphs*, i.e., graphs with multiple parallel edges connecting vertices.

A search for greedy or hungry spiders on the https://arxiv.org[2] turned up no results, and as far as we know, there are no mathematical papers studying the hungry spiders game. However, hungry spiders fits a larger category of graph problems called edge deletion problems, which ask to transform an input graph into a graph in a given, specific class of graphs by deleting a minimum number of edges. There is a wealth of literature on edge deletion problems, and to start studying their complexity, we recommend reading Yannakakis' article from 1981 and any of the many articles citing it on the SIAM website [10].

We remark that there are countless variations or additions one could make to the game. In Blyts' game, the player gets additional move types as the levels get more challenging. For instance, the player can place a decoy fly that may lure the spiders away from the real flies for a time period, or the player can delete a vertex and all its incident edges from the graph. For further study, consider the following list of possibilities or brainstorm your own:

- Change the spiders' algorithm from a greedy rule or play against another person.
- Allow the player to delete one or more vertices as well as deleting an edge in each turn.
- Place decoy flies to lure the spiders away from the real ones.
- Use a digraph as the web in which spiders must move in the direction of the arcs.
- Allow the flies the special option to teleport to a different vertex one time per round.
- Rather than alternating turns, a coin-flip or a spinner could decide who gets to go next. In this case, winning the game on a particular graph would also depend on chance.

4 (t, r) Broadcast Domination

A known generalization of the process of graph domination is known as (t, r) broadcast domination, and it is defined using the following analogy. Imagine some vertices of a grid graph represent houses, some represent cell phone towers, and we are allowed to freely choose which of the vertices are houses and which are the towers. The towers all emit some signal of a known strength. The goal of this game

---

[2] This is a repository of articles used by mathematicians to share their most recent results in a broad range of mathematical areas.

is to ensure that we can place towers so that all houses receive a minimum amount of necessary signal using as few towers as possible.

In particular, we can say that each tower gives itself signal strength t (where t is a positive integer), and its signal strength decays linearly outward from vertex to vertex as it traverses the edges of the graph. More specifically, if a tower is located on vertex u, then each vertex $v \in V(G)$ receives signal strength $t - \text{dist}(u, v)$ from that tower. If there are multiple towers whose signal reaches a single house, we add those signals together to determine the amount of signal the house receives. Given a graph G and positive integers t and r, a (t, r) *broadcast dominating set* of G is a set of vertices on which towers (with signal t) can be placed so that every vertex in $V(G)$ receives a total signal of at least r. The (t, r) *broadcast domination number of G*, denoted $\gamma_{t,r}(G)$, is the minimum size of a (t, r) broadcast dominating set in G. Intuitively, this number is the fewest number of towers needed in G to ensure that every vertex receives sufficient reception.

It is worth pointing out that $(2, 1)$ broadcast domination is equivalent to the standard notion of graph domination where each tower's signal only reaches that tower's direct neighbor, and we only require that each vertex on the graph either is a tower or is a house receiving signal from at least one tower. We refer the readers interested in a comprehensive study in (classical) graph domination to consult the text by Haynes, Hedetniemi, and Slater [7].

Figure 8 illustrates an example of a $(3, 2)$ broadcast dominating set of a 3×5 grid graph. In this example, note that the vertices of distance 2 from any tower receive low signals (of strength 1) from each tower. However, the low signals from each tower add together to give these vertices sufficient reception.

The study of (t, r) broadcast domination number of grid graphs originated with the work of Blessing, Insko, Johnson, and Mauretour in 2015 [1]. In this paper, the authors determined the broadcast domination number for various small values of t and r for $2 \times n$, $3 \times n$, $4 \times n$, and $5 \times n$ grid graphs. Consider the following challenge problems for an opportunity to practice broadcast domination on grid graphs.

Challenge Problem 10 Show that the vertices of the 3×5 grid graph cannot all receive minimum signal $r = 2$ from only two towers with strength $t = 3$.

Fig. 8 The 3×5 grid graph with $t = 3$ towers (shown in red) placed at positions $(1, 1)$, $(3, 3)$, and $(1, 5)$. The signal receptions are shown on the vertices

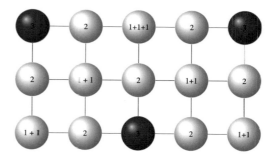

Challenge Problem 11 Describe how to find a minimum $(2, 2)$ dominating set on a $3 \times n$ grid graph for each $n \geq 3$.

Challenge Problem 12 Describe how to find a minimum $(2, 2)$ dominating set on a $4 \times n$ grid graph for each $n \geq 3$. [*Hint: it is known that the* $(2, 2)$ *domination number of* $G_{4,n}$ *is* $2n - \left\lceil \frac{n-6}{4} \right\rceil$.]

While exact formulas for some (t, r) broadcast domination numbers exist, extending this work is still an open area of research, leading to the following research project (originally posed in 2020 in [6]).

> **Research Project 8** Find formulas for the (t, r) broadcast domination number for grid graphs $G_{m,n}$ with $m \in \{2, 3, 4, 5\}$, $n \in \mathbb{N}$, and (t, r) different from $(2, 2)$, $(3, 1)$, $(3, 2)$, $(3, 3)$.

We can even think of converting these grid graphs into cylinders by "gluing" the right and left ends of the grids as follows: for every row, simply connect the rightmost and leftmost vertices of that row by an edge. Now, we can consider the following research project.

> **Research Project 9** Given a grid graph $G_{m,n}$ and values t, r, connect the vertices in the rightmost column of $G_{m,n}$ to those in the leftmost column with edges, and find formulas for the (t, r) broadcast domination number of the resulting graph.

Another potentially fruitful area of research is to find patterns for efficient broadcast domination in graph families which are altogether different from grid graphs. Consider the following challenge problem to practice thinking about broadcast domination for other graph families.

Challenge Problem 13 Let $t > r \geq 1$. What is the (t, r) broadcast domination number of a complete graph on n vertices?

There are many more families for which nothing is known about broadcast domination. The following research project provides an opportunity for great discovery.

> **Research Project 10** Determine the (t, r) broadcast domination number of various families of graphs. As a starting point, consider the Petersen graph, cycles, trees, and complete graphs (see Sect. 6 for definitions of these graph families).

It is also possible to study broadcast domination in the context of various common graph operations. The following challenge problem highlights how the (t, r) broadcast domination number may change after performing these operations.

Challenge Problem 14 Suppose t and r are integers with $1 \leq r < t$. Construct a graph G for which deletion of any edge results in an increase of the (t, r) broadcast domination number. Construct a graph G on which deletion of any edge does not affect the broadcast domination number.

Further study of the intricacies of vertex and edge deletion (and other graph operations) may lead to interesting findings and a greater understanding of the changes in (t, r) broadcast domination number as one transitions from one graph to another using those operations. This leads to the following research project.

> **Research Project 11** Study the effects of various graph operations on (t, r) broadcast domination number of a graph. How does this number change when an edge or a vertex is deleted? How does the (t, r) broadcast domination number of a graph G compare to that of its complement $\overline{G}$? How about when an edge is subdivided (i.e., we add a vertex x to $V(G)$ and replace an edge (u, v) with edges $(u, x), (x, v)$)?

A process that has just been proposed within the past year is that of *directed broadcast domination* [5]. Given a graph G, orient each of its edges to create an orientation $\vec{G}$ of G. Now, the process of directed broadcast domination works in $\vec{G}$ much like in G, except we only allow signal to travel in one direction along the arcs of $\vec{G}$.

Note that the directed (t, r) broadcast domination number of $\vec{G}$ may change depending on which orientation $\vec{G}$ we choose. To this end, define the (t, r) *directed broadcast domination interval* of a graph G, denoted (t, r) DBDI(G), to be the smallest closed interval that contains all possible (t, r) directed broadcast domination numbers of $\vec{G}$ taken over all orientations of G.

Challenge Problem 15 What is the $(3, 2)$ directed broadcast domination interval of the cycle on six vertices?

The following project in directed broadcast domination explores the bounds of this interval for various families of graphs.

> **Research Project 12** Given values of t and r, what are the maximum and minimum values in the (t, r) DBDI of various families of graphs?

Graphs may also be studied within a probabilistic context using randomness. The following problem highlights (t, r) broadcast domination as a problem in probability.

Challenge Problem 16 Let $t = r = 2$. For each $x \in \{1, 2, 3, 4, 5\}$, what is the probability that a randomly selected set of vertices of size x is a (t, r) broadcast dominating set of a path on 5 vertices?

One can imagine a real-life application of random (t, r) broadcast domination. For example, in some computer networks, the network "hosts" may be shuffled around randomly within a group from time to time with some arrangements more optimal than others. This creates a very applicable area of research.

> **Research Project 13** Given an arbitrary graph G and fixed values t and r, what is the probability that a randomly selected set of vertices is a (t, r) broadcast dominating set?

Lastly, there are many ways in which broadcast domination can be altered to produce entirely new fields of research. One potential direction for study is to alter the way in which signal decays over the edges of a graph. For example, consider the following research project.

> **Research Project 14** Study (t, r) broadcast domination when signal decays nonlinearly over edges. One example of this is to reduce the signal by half across every edge. Note that when $r = 1$, this process resembles *exponential domination*, described in detail in [4].

5 Further Investigation

As we have seen, many problems in analyzing games on graphs, such as the ones we presented in this chapter, can be pondered with relatively little mathematics background. This makes games on graphs an incredibly accessible area of study for students who are curious and willing to learn and play. Yet we have only presented very few of the many ways in which these games can be studied. In fact, with more background in probability, combinatorics, linear algebra, and abstract algebra, students could continue exploring the games we presented with new techniques. Such investigations could easily lead to research projects for undergraduate theses or even lead to Ph.D. dissertation projects in mathematics. We encourage students to continue in their mathematical journey, keep asking questions, but more importantly have fun while playing games!

6 A Short Primer on Graph Theory

We begin by introducing some needed mathematical background and notation. A *graph* G is defined by a set of *vertices* (dots), denoted by $V(G)$, and a set of *edges*, denoted by $E(G)$, where an edge is drawn as a line segment connecting two vertices. If the graph G is clear from context, we write V and E for the vertex and edge sets, respectively. If an edge e connects vertex u to vertex v, we say that u and v are *endpoints* of e, and we often write $e = uv$. Furthermore, every edge is said to be *incident* to its endpoints. When we say a graph is *finite*, we mean that the number of vertices is finite, and when we say the graph is *simple*, we mean that every edge must have two distinct endpoints (no loops), and there can be at most one edge incident to any pair of distinct vertices (no parallel edges). Throughout this chapter, we will always assume that G is a finite simple graph. Table 1 illustrates some common families of finite, simple graphs.

Note that the $m \times n$ *grid graph* $G_{m,n}$ has vertices arranged into an array consisting of m rows and n columns. Therefore, the grid graph as drawn in Table 1 is denoted as $G_{2,3}$. Another simple graph that is particularly important in graph theory is called the *Petersen graph*. The Petersen graph has 10 vertices and 15 edges and is illustrated in Fig. 9.

If a graph G is not a member of a well-defined family, or if it does not already have a special name, we can still describe G using a variety of terms and concepts from graph theory. For instance, we say that a graph is *connected* if, given any two vertices u and v, we can find a way to "walk" from u to v by traveling along the edges of the graph to get from one vertex to another. Note that the graphs shown in Table 1 and Fig. 9 are indeed connected. We say a graph contains a *cycle of length n* if we can find a vertex u and a "walk" along n distinct edges of the graph that allows us to begin and end at the vertex u. Note that the *wheel graph* W_6 illustrated in Table 1 contains many cycles of length three. Graphs that are connected and contain no cycles are called *trees*. The *path graph* P_n and the *star graph* S_n are the only graphs listed in Table 1 that are trees for each integer $n \geq 1$. Do note that the grid

Table 1 Each row describes a family of finite, simple graphs and depicts an example on 6 vertices.

Name	Notation	Vertices	Edges	Example (on 6 vertices)
Path	P_n	n	$n-1$	
Cycle	C_n	n	n	
Star	S_n	n	$n-1$	
Wheel	W_n	n	$2(n-1)$	
Complete graph	K_n	n	$\dfrac{n(n-1)}{2}$	
Grid graph	$G_{m,n}$	mn	$(m-1)n+(n-1)m$	

Fig. 9 The Petersen graph

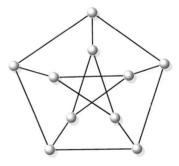

graph $G_{m,n}$ is a tree whenever $m = 1$ or $n = 1$, and this happens because then $G_{1,n}$ is a path graph on n vertices and $G_{m,1}$ is a path graph on m vertices.

Graph connectedness is a binary property. In other words, for a graph to be connected, there must exist at least one way to get from any vertex of the graph to any other by traversing ("walking") along the edges of the graph. If such is not the case for merely one pair of vertices, then the graph is not connected. However, there may actually be multiple ways to do a walk. In light of this, we might want to be as efficient as possible in our "walking" and only traverse the minimum number of edges required to get from vertex u to vertex v. This is the concept of *distance* within a graph, which is defined as the minimum number of edges we can "walk" along to get from vertex u to vertex v. The distance between u and v is denoted by dist(u, v). By convention, if we cannot find a "walk" that allows us to go from u to v, then dist$(u, v) = \infty$.

There are many other definitions that allow us to describe the structure of a graph. We begin with the following. Two vertices that are both incident to the same edge are said to be *adjacent* and are often called *neighboring* vertices or simply *neighbors* for short. The *degree* of a vertex v in a simple graph, denoted deg(v), is the number of vertices in $V(G)$ that are adjacent to v. The *maximum degree* of a graph G, denoted $\Delta(G)$, is the maximum value of deg(v) taken over all vertices $v \in V(G)$. Alternatively, the minimum such value, denoted $\delta(G)$, is the *minimum degree* of G. For example, the Petersen graph G (in Fig. 9) satisfies that deg$(v) = 3$ for every vertex v, so that means that $\delta(G) = \Delta(G) = 3$, whereas the star graph S_6 has 5 vertices with degree one and one vertex with degree five, hence $\delta(S_6) = 1$ and $\Delta(S_6) = 5$. In general, $\delta(S_n) = 1$ and $\Delta(S_n) = n - 1$ for $n \geq 1$.

There are also many useful operations that we can perform on graphs. To *delete an edge* from a graph G, we simply remove the edge from $E(G)$. To *delete a vertex* v from a graph G, we remove v from $V(G)$, and we remove from $E(G)$ any edge that is incident to v. If $S \subseteq V(G) \cup E(G)$, the graph obtained from G by removing the vertices and edges in S is denoted by $G - S$. For example, Fig. 10 illustrates taking the complete graph on six vertices and deleting a vertex along with the edges incident to it. The *complement* of a simple graph G, denoted $\overline{G} = K_{|V(G)|} - E(G)$ the graph obtained by deleting all the edges of G from the complete graph $K_{|V(G)|}$.

delete v

Fig. 10 Deleting the vertex v from K_6

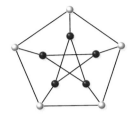
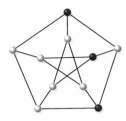

Fig. 11 In each subfigure, S is made up of the colored vertices. Left: a set S, which does not dominate the Petersen graph. Center: a set S, which dominates the Petersen graph. Right: a set S, which dominates the Petersen Graph with a minimum number of vertices

Fig. 12 An example of a digraph, which is an orientation of P_6

Sometimes, we are interested in certain subsets of $V(G)$ with special properties. Given a graph G, a dominating set of G is a set of vertices such that any vertex in the graph is either in the set or adjacent to a vertex in the set.

When S is a dominating set of graph G, we say that S *dominates* G. Note that for any graph, selecting every vertex of the graph will always lead to a dominating set. We might then ask how much smaller we could make the size of a dominating set. That is the idea behind the *domination number* of a graph. The *domination number* of G, $\gamma(G)$ is the minimum size (order) of a dominating set of G. For example, Fig. 11 illustrates a subset of vertices of the Petersen graph G which does not dominate G, another which does, and a third which does so with a minimum number of vertices possible.

In a *directed graph (or digraph)*, the edges (also called *arcs*) are ordered pairs of vertices. This gives a direction to the edges because in a digraph Γ, it is possible that $(u, v) \in E(\Gamma)$, yet $(v, u) \notin E(\Gamma)$. We can draw the arc (u, v) as an arrow that points from u to v. In this case, u is called an *in-neighbor* of v and v is called an *out-neighbor* of u. In Fig. 12, we orient the edges of the path graph P_6 to create a digraph. If Γ is a digraph and $v \in V(\Gamma)$, the *in-degree* of v, denoted $\deg^-(v)$, is the number of in-neighbors of v in $V(\Gamma)$. Likewise, the *out-degree* of v, denoted $\deg^+(v)$, is the number of out-neighbors of v in $V(\Gamma)$. If we start with a simple graph G and replace each edge $uv \in E(G)$ with either the arc (u, v) or the arc (v, u), the resulting digraph is called an *orientation* of G. Note that Fig. 12 is an orientation of the path graph P_6.

References

1. David Blessing, Katie Johnson, Christie Mauretour, and Erik Insko. On (t, r) broadcast domination numbers of grids. *Discrete Appl. Math.*, 187:19–40, 2015.
2. Joshua Carlson, Robin Eagleton, Jesse Geneson, John Petrucci, Carolyn Reinhart, and Preetul Sen. The damage throttling number of a graph. Under review.

3. Danielle Cox and Asiyeh Sanaei. The damage number of a graph. *Australas. J. Combin.*, 75:1–16, 2019.
4. Peter Dankelmann, David Day, David Erwin, Simon Mukwembi, and Henda Swart. Domination with exponential decay. *Discrete Math.*, 309(19):5877–5883, 2009.
5. Pamela E. Harris, Peter Hollander, and Erik Insko. The (t,r) broadcast domination number of directed graphs. in-progress, 2021.
6. Pamela E. Harris, Erik Insko, and Katie Johnson. Projects in (t,r) broadcast domination.
7. Teresa W. Haynes, Stephen T. Hedetniemi, and Peter J. Slater. *Fundamentals of domination in graphs*, volume 208 of *Monographs and Textbooks in Pure and Applied Mathematics*. Marcel Dekker, Inc., New York, 1998.
8. Richard Nowakowski and Peter Winkler. Vertex-to-vertex pursuit in a graph. *Discrete Math.*, 43(2-3):235–239, 1983.
9. A. Quilliot. Some results about pursuit games on metric spaces obtained through graph theory techniques. *European J. Combin.*, 7(1):55–66, 1986.
10. Mihalis Yannakakis. Edge-deletion problems. *SIAM J. Comput.*, 10(2):297–309, 1981.

Mathematics for Sustainable Humanity: Population, Climate, Energy, Economy, Policy, and Social Justice

Jue Wang

Abstract

Sustainable development means meeting our needs without compromising the ability of future generations to meet their own needs. Human activities have significantly altered the ecosystems, causing environmental degradation, loss of biodiversity, and climate change. Accompanied with the rapid economic growth is the increasing consumption of natural resources and rising social inequalities. The environment, economy, and society must function together and support each other for a sustainable humanity. Mathematics is essential in identifying and analyzing the challenges. It helps us understand the impact and assess the risks. It allows us to make predictions to better inform science and public policy. Through a sequence of sustainability and mathematical concepts, exercises, and projects, the reader is brought along on this journey and confronts global challenges to develop an understanding of the complex environmental, economic, and sociocultural interlinkages and to equip themselves to cooperate actively in the present and support sustainable development in the future.

Suggested Prerequisites *None.*

J. Wang (✉)
Union College, Schenectady, NY, USA
e-mail: wangj@union.edu

© The Author(s), under exclusive license to Springer Nature Switzerland AG 2022   21
E. E. Goldwyn et al. (eds.), *Mathematics Research for the Beginning Student,*
Volume 1, Foundations for Undergraduate Research in Mathematics,
https://doi.org/10.1007/978-3-031-08560-4_2

1   Introduction

Every investigative journey begins with guiding questions, such as:

- Can economic growth last?
- How big is the inequality gap?
- Is renewable energy growing fast enough?
- Is our climate headed for a mathematical tipping point?
- How long can our biological systems remain diverse and productive?
- How much CO_2 emissions result from consumption of bottled water and paper coffee cups on your school campus each day?

These are sustainability questions. The concept of sustainability is born out of the desire of humanity to continue to exist on planet Earth. It has evolved far beyond the ecological context. Today, it addresses the idea of a global society founded on respect for nature, universal human rights, economic justice, and a culture of peace [9]. **Sustainability** is the ability of humanity to meet the needs of the present generation without compromising the ability of future generations to meet their own needs [6]. It is of vital importance and will affect everyone.

Sustainability is a highly interdisciplinary subject. It draws on diverse perspectives, including scientific, technological, political, ethical, and religious. The three dimensions (or pillars) of sustainability are environmental protection, economic development, and social well-being, illustrated in Fig. 1. These three elements form an integrated, cohesively interacting system.

Sustainability is becoming increasingly urgent. The immense and rising needs of humanity have posed a threatening impact on the climate, biodiversity, and freshwater supplies. The global assessment report on biodiversity and ecosystem services, released in 2019 by the United Nations' Intergovernmental Science-Policy Platform on Biodiversity and Ecosystem Services (IPBES), assessed the status of biodiversity and human impact on the environment [27]. Over the past 50 years, the Earth's biodiversity has suffered a catastrophic decline, and its ecosystems have been degraded at rates unprecedented in human history.

The large net gains in economic development are achieved at substantial costs of environmental destruction and loss of diversity of life on the Earth. This inevitably

Fig. 1 The three dimensions of sustainability

results in harmful consequences to human well-being, growing inequities, and social conflict. Both the society and economy must be supported by a functioning environment. Reversing the degradation of nature while meeting the increasing demands will involve significant changes and enactment in policy, practice, and institution. This is a fundamental challenge that we must collectively face and tackle.

Mathematics plays an important role in identifying, measuring, and analyzing the issues. Mathematical models not only help us understand the natural phenomena but also allow us to describe the changing process and predict the future. Mathematics provides key tools to evaluate the evidence, quantify the effects, assess the risks, and suggest hypotheses to inform science and public policy.

This chapter will introduce the essential concepts of sustainability and help you gain the mathematical knowledge, so that you can apply it to address the challenges facing our planet. Five real-world projects will lead you to examine the impact brought to our society, economy, climate, and natural resources. Each project is independent. Collectively, they cover the multifaceted nature of sustainability. In addition, you will work with the latest wide range of data, create powerful visualizations, interpret and draw inferences from functions, equations, models, and graphs, as well as developing an understanding of how data can be used to make different arguments. You will gain insights from the mathematics and science of resilience and fold them into policy decisions.

This chapter aims to guide you to make a connection between mathematics and real-life issues we care about, to engage in critical thinking and quantitative reasoning, to develop step-by-step problem solving skills, to use the knowledge to shape competency and reach justified decisions, and to empower yourself to become an active citizen.

2 Quantifying Change

How big is the CO_2 emission? What is the significance of the value? This naturally leads to the question of metrics—how to measure and estimate?

2.1 Absolute and Relative Change and Rate of Change

Ratios and percentages are commonly used to measure the level of growth or decay. They are unit-free measures. For example, the energy payback ratio of an electrical power generating plant is the ratio of the electrical energy produced to the energy costs of building, operating, and decommissioning the facility.

Energy payback ratio = electrical energy generated/energy consumed by facility

Exercise 1 A natural gas-fired electrical plant produces 13,600,000 gigajoules of electricity per year at a cost of 3,200,000 gigajoules of energy per year (giga $= 10^9$). What is its energy payback ratio of this facility?

Coal and gas-fired power plants have energy payback ratios in the range of 3–5, while wind-driven power plants have energy payback ratios in the range of 18–34 [5].

Percentage is a type of ratio. For example, the monthly average concentration of CO_2 in the atmosphere in April 2021 and April 2020 was 419.05 ppm and 416.45 ppm, respectively [41], with ppm standing for parts per million by volume. We can express 419.05 ppm as a percentage:

$$\frac{419.05 \ m^3 \ [CO_2]}{1,000,000 \ m^3 \ [\text{atmosphere}]} = \frac{0.041905}{100} = 0.041905\%$$

The April CO_2 concentration has increased in one year by

$$419.05 \ \text{ppm} - 416.45 \ \text{ppm} = 2.60 \ \text{ppm}$$

This is the **total change**, or **absolute change**, which is the actual increase or decrease amount of a quantity: $Q_1 - Q_0$. Another way to quantify change is to use the **percentage change**, or **relative change**, which is the percentage by which a quantity increases or decreases: $(Q_1 - Q_0)/Q_0$ for nonzero Q_0.

In the above example, the one-year percentage increase of CO_2 concentration is

$$\frac{(419.05 - 416.45) \ \text{ppm}}{416.45 \ \text{ppm}} = \frac{2.60 \ \text{ppm}}{416.45 \ \text{ppm}} \approx 0.0062 = 0.62\%$$

We can further calculate the **growth factor** as $1 + percentage \ change$, that is,

$$1 + 0.0062 = 1.0062$$

Observe: What do you notice?

This is exactly the ratio between the two amounts: Q_1/Q_0 ($Q_0 \neq 0$). This defines the **growth factor** if $Q_1 > Q_0$, or if the percentage change is positive (increase), and the **decay factor** if $Q_1 < Q_0$, or if the percentage change is negative (decrease).

Now that we have seen different ways change can be measured, can we get a sense of how fast something is changing? For example, suppose a population grew by 10%, is it growing fast or slow? That depends on how long it took. This rate at which something changes can be quantified by the average **rate of change**, the amount of change over one time unit: $(Q_1 - Q_0)/\triangle t$. As opposed to ratio and percentage, this measure has units.

Exercise 2 What is the growth or decay factor and annual percentage change if a population doubles each year? Triples each year?

Exercise 3 The table below gives the global temperature anomaly and September Arctic sea ice extent in 1980 and 2020.

	1980	2020
Global temperature anomaly	0.23 °C	0.92 °C
September Arctic sea ice extent	7.67 million square kilometers	3.93 million square kilometers

For the temperature anomaly and Arctic sea ice extent respectively,

(a) What is the growth/decay factor? Keep 4 decimal places.
(b) What is the total change?
(c) What is the percentage change? Specify increase or decrease.
(d) What is the average rate of change?

Include appropriate units when applicable.

> **Think** What will happen to a quantity if the total change (absolute change) remains constant over equal periods of time? What if the percentage change (relative change) remains constant? What will happen to the rate of change?

With your thoughts in mind, let us explore these relationships and make connections.

2.2 Linear and Exponential Change

Suppose a population group A starts with 100 people. It grows by 22 people during any 2-year period. This means that the total change for population A is the same. The population size over time is listed in the following table. Do a quick check yourself.

Exercise 4 Plot the data. What do you observe? What would happen if the trend continued? Find the average rate of change over each 2-year period (in units of people per year). What do you notice in the rate of change?

	Current	2 years	4 years	6 years	8 years	10 years
Population A	100	122	144	166	188	210

	0–2 years	2–4 years	4–6 years	6–8 years	8–10 years
Population A					

Through this exercise, we see that if the total change remains constant over equal periods of time, then the quantity follows a **linear growth (or decay)** and the rate of change is a constant. We can express the change using a linear function and project future values. Let x be the number of years since the starting year, and let y be the corresponding number of people. Then, population A follows a linear function

$$y = 100 + 11x,$$

where 100 is the starting value and 11 is the constant rate of change. After 10 years, we would have $y = 210$, as previously obtained in the table.

Suppose another population group B also starts with 100 people. It grows by 20% during any 2-year period. This means that the percentage change for population B is the same. The population size over time is listed below, with values rounded to the nearest integers. You may calculate yourself.

	Current	2 years	4 years	6 years	8 years	10 years
Population B	100	120	144	173	208	250

Exercise 5 Plot the data. What do you observe? What would happen if the trend continued? Calculate the growth factor for population B. Find the average rate of change over each 2-year period. What do you notice in the rate of change?

	0–2 years	2–4 years	4–6 years	6–8 years	8–10 years
Population B					

Through this exercise, we see that if the percentage change remains constant over equal periods of time, then the quantity follows an **exponential growth (or decay)** and the growth (or decay) factor is a constant, but the rate of change increases. We can express the change using an exponential function for Population B,

$$y = 100 \cdot (1.2)^{x/2}$$

> **Think** Can you explain where the 1.2 comes from and why the exponent is $x/2$?

Now, let us put the two populations together for comparison.

	Current	2 years	4 years	6 years	8 years	10 years
Population A	100	122	144	166	188	210
Population B	100	120	144	173	208	250

Rate of change:

	0–2 years	2–4 years	4–6 years	6–8 years	8–10 years
Population A					
Population B					

Exercise 6 What do you notice in the data? Plot the data together. What do you observe? Predict the population sizes for A and B after 50 years using the linear and exponential functions, respectively.

To summarize, a **linear function** takes the form

$$y = mx + b,$$

where m represents the constant total change of y for each 1-unit increase in x, and b represents the initial amount of y at $x = 0$. The graph is a straight line with slope m and y-intercept b. An **exponential function** takes the form

$$y = b \cdot (c)^{x/n},$$

where b stands for the initial amount, c is the constant growth/decay factor, and n is the time period for c. Exponential growth is much faster than linear growth.

Exercise 7 Which of the following scenarios describes a linearly changing quantity? Which describes an exponentially changing quantity? Give a brief reason.

(a) The September Arctic sea ice is declining at a rate of 13% per decade.
(b) The New England herring gull population has been doubling every 15 years.
(c) The rain was falling about two inches every hour during hurricane Florence.

Did You Know? Since the dawn of civilization, the wild mammal biomass has fallen by 82%; natural ecosystems have lost about half their area. A million species are currently threatened with extinction, including more than 40% of amphibians, over a third of marine mammals, and nearly a third of reef-forming corals [27].

2.3   Measuring and Estimating

Carbon footprint is the amount of emissions caused by an individual, organization, or activity, expressed as carbon dioxide equivalent. US greenhouse gas emissions totaled 6.6 billion metric tons of carbon dioxide equivalents in 2019 [17]. Carbon dioxide (CO_2) accounted for 80% of the total, among others including methane (CH_4), nitrous oxide (N_2O), and fluorinated gases. The sources of greenhouse gas emissions by economic sector in the USA are illustrated in Fig. 2. Transportation is the largest sector, accounting for 29% of greenhouse gas emissions in 2019. Light-duty vehicles contributed to 58% of the transportation sector emissions.

The average carbon footprint for a person in the USA was 16.1 tons in 2019, one of the highest in the world. Globally, this average was 4.7 tons in 2019 [25]. Lowering individual carbon footprints from 16 to 5 tons does not happen overnight. What percentage decrease is that? By making small changes to our actions, we can start making a big difference. How do we calculate the carbon footprint? How can one estimate the size, value, and significance of personal carbon emissions?

Exercise 8   How Big Is Your Breakfast Carbon Footprint?

In this simple exercise, you will take your everyday breakfast food items and estimate the associated carbon footprints. We use the following formula [10]:

Greenhouse Gas Emissions $= D \times W \times EF$

$D =$ The distance your shipment has traveled
$W =$ The weight of the shipment
$EF =$ The mode's specific emissions factor

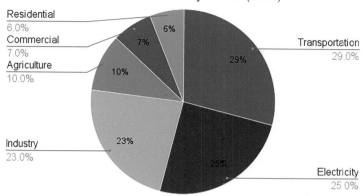

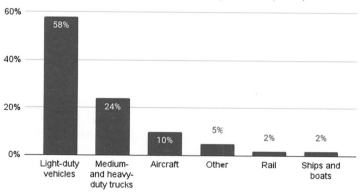

Fig. 2 Top: 2019 US greenhouse gas emissions by economic sector, in pie chart. Bottom: US transportation sector greenhouse gas emissions by source, in column chart. Totals may not add to 100% due to rounding. "Other" sources include buses, motorcycles, pipelines, and lubricants. Data source: US Environmental Protection Agency (EPA) [18]

The average freight truck in the USA emits 161.8 grams of CO_2 per short ton-mile [11]. This is the emission factor (EF).

Note: the **short ton** is commonly used in the USA, known as "ton" or US ton.

1 US ton = 2,000 lb (or 907.2 kg)

It is distinguished from the **metric ton**, known as "tonne" or the long ton.

(continued)

1 tonne = 1,000 kg (or 2,204.6 lb)

1 US ton = 0.9 tonne,   1 tonne = 1.1 US ton

Only three countries—Liberia, Myanmar (formerly, Burma), and the USA—have not adopted the International System of Units (SI, or metric system) as their official system of weights and measures. Liberia uses the US customary system (USCS), and Myanmar uses the UK Imperial system.

Two examples of the **unit factor method** for unit conversions are demonstrated below. You may refer to [2] for more examples and conversion factors between the metric and US systems.

$$1 \text{ ton} = 1 \text{ ton} \times \frac{2000 \text{ lb}}{1 \text{ ton}} \times \frac{0.4536 \text{ kg}}{1 \text{ lb}} = 907.2 \text{ kg}$$

The second example demonstrates how to put the volume of fresh surface water on the Earth [54] into perspective:

$$104,590 \text{ km}^3 = 104,590 \text{ km}^3 \times \frac{1 \text{ Lake Superior}}{12,000 \text{ km}^3} = 8.7 \text{ Lake Superiors}$$

Use the following table to guide you through the calculations. Add more rows if needed. For simplicity, we will only look at the fraction of carbon footprint associated with transportation. Identify your breakfast item, its weight, and origin (printed on the package) and obtain the mileage from the origin location to your home (or a local store) with Google map. Then, use the greenhouse gas emissions formula to calculate the carbon cost. Include all units in your calculations and be careful with unit conversions.

Breakfast food item	Origin	Miles	Weight (g)	Carbon cost (g of CO_2)

Here is an example for half-gallon milk, comparing two options in a local store.

	Origin	Miles	Weight (g)	Carbon Cost	
Maple Hill Organic Milk	Kinderhook, NY	40	1892.7	13.47	g of CO_2
Horizon Organic Milk	Broomfield, CO	1792	1892.7	603.66	g of CO_2

Tip We human beings are not always able to grasp very large or very small numbers or units of measurement that do not relate to a familiar scale. If we can express a measurement in terms of more familiar quantities, it can help make sense. Keep this in mind when you try to understand or express the significance of a quantity.

To make sense of the amount of CO_2, let us convert gram to volume cm^3, and determine how many volleyballs can be filled. The calculation is similar to the example of unit factor method for visualizing the amount of fresh surface water on the Earth. The molecular weight of CO_2 is 44 g per mol. One mol of ideal gas occupies 22.4 liters at 1 atmosphere and room temperature. The diameter of a volleyball is 8.15 in, or 20.7 cm, so the radius is 10.35 cm.

What is the volume of the volleyball? How many volleyballs can be filled by the CO_2 emissions from merely transporting your breakfast food items?

Breakfast food item	Origin	Miles	Weight (g)	Can fill # volleyballs

Now, we can get a better handle in comparing the carbon footprint of these two milk options.

	Origin	Miles	Weight (g)	Can Fill	
Maple Hill Organic Milk	Kinderhook, NY	40	1892.7	1.5	volleyballs
Horizon Organic Milk	Broomfield, CO	1792	1892.7	66	volleyballs

An infographic visualization is shown in Fig. 3, which is one simple yet powerful way to present data. You have just reasoned yourself one of the many motivations for buying local and why it is better for the planet.

For more information about carbon footprint reductions, visit EPA [16]: carbon footprint calculator and greenhouse gas equivalencies calculator.

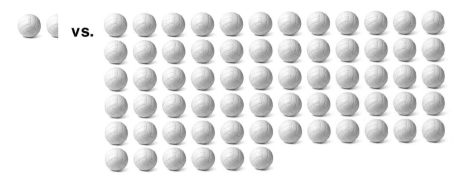

Fig. 3 Volleyballs filled by CO_2 emissions from transporting half-gallon local vs. nonlocal milk

Did You Know? The global disposable paper cups market reached a volume of 263 billion units in 2020 [15]—enough to go to the moon and back 30 times if placed end to end! Over 150 million disposable paper cups are thrown away in the USA each day—enough to circle the equator once in less than 3 days! The sheer number of cups is clearly staggering, but the real controversy is that these cups are not being recycled—small paper cups, BIG plastic problem!

Most paper cups for consuming coffee, tea, soups, and soft drinks are lined with a thin layer of plastic—polyethylene (PE). This not only prevents leaking but also makes the cup non-recyclable, which comes as a surprise to many. It takes about 30 years for such a cup to decompose [60]. These cups must go to special recycling facilities that can remove the plastic laminate from the exterior of the cup. Few recycling facilities can process this. It is estimated that the manufacture of every 4 paper cups results in one lb of CO_2 emissions. Just consuming one cup of coffee in a disposable paper cup every day will end up creating about 23 lbs of waste in one year.

It is not just the paper cup that is an environmental nightmare. Other plastic-coated paper products include paper plates, milk cartons, and juice cartons. Studies show conclusively that microplastic fragments (less than 5 millimeters in length, or about the size of a sesame seed) shed from all plastic-coated samples in finished compost, posing a significant risk to soil, freshwater, wildlife, and humans [3]. Microplastic was found in the gut of every single sea turtle examined from all seven marine turtle species, across the Atlantic, Pacific, and Mediterranean [7].

Plastic is everywhere. Plastic bottles and bottle caps rank as the third and fourth most collected plastic trash items in the Ocean Conservancy's annual beach cleanups in more than 100 countries [34]. Plastic bottles take at least 450 years to decompose. Only 1 in 5 plastic bottles gets recycled. The energy required to produce and transport plastic water bottles could fuel 1.5 million cars for a year [60].

Exercise 9 How much CO_2 emissions result from consumption of bottled water and paper coffee cups on your school campus each day?

Suppose 200 cups of hot coffee and 125 bottles of water are sold on your school campus each day (for a 2,200-student campus). You may obtain better estimates of these numbers based on your school size. Similar to the method from Exercise 8, estimate the amount of CO_2 emissions from consumption of paper coffee cups and bottled water on your school campus, assuming that bottled water transportation is within 300 miles. For simplicity, you may consider only the fraction of CO_2 emissions from paper coffee cup manufacture (0.25 lbs per cup) and bottled water transportation (300 miles distance).

Photographer Chris Jordan uses large format photography to portray the extent of our mass consumptions. His images confront the enormous power of humanity's collective will, calling for cultural self-inquiry and self-awareness. Among many intriguing works, his "Paper Cups" (5 feet by 8 feet, 2008) [31] depicts 410,000 paper cups, equal to the number of disposable hot-beverage paper cups used in the USA every 15 min (in 2008). The number of paper cup consumptions has more than tripled since 2008. His "Blue" (5 feet by 5.8 feet, 2015) [32] depicts 78,000 plastic water bottles, equal to 1/10, 000th of the estimated number of people in the world who lack access to safe drinking water.

> **Takeaway** Bring your own mug. Refill your water bottle. Small actions can lead to a huge impact!

3   Population Growth and Ecological Footprint

The world population has grown rapidly, from approximately 2.6 billion in 1950 to 7.8 billion in 2020. The increase in population and resource consumption has put an enormous amount of pressure on our planet, making it harder to sustain on a planet that is after all finite. Humanity needs to embrace some significant actions to shrink our ecological footprints.

What is an ecological footprint? The **ecological footprint** is the only metric that measures how much nature we have (supply) and how much nature we use (demand) [22]. It not only helps us understand our personal impact on the planet but also gives us a better sense of the ecological impact of cities, nations, and the global community.

On the demand side, the ecological footprint measures the ecological assets required to produce the natural resources consumed (including food, fiber, livestock, fish, timber, and space for urban infrastructure) and to absorb waste. On the supply side, the **biocapacity** represents the productivity of its ecological assets (including cropland, grazing land, forest land, fishing grounds, built-up land, and carbon uptake land). This supply–demand balance is vital for sustainable development.

Research Project 1

In this project, you will explore the demand on ecological resources and supply available and gain an insight of how sustainably we are living.

Part 1. Population Projection

Obtain the latest population data (xlsx) from the United Nations, Department of Economic and Social Affairs [51]. The data list the total population by region, subregion, and country, annually from 1950 to present (in thousands). You may examine the world population as a whole or choose a country, and work with Excel (tutorial [21]) or Google Sheets (tutorial [26]). Here, we will use the population of the USA as an example: from 158,804.397 in 1950 to 331,002.647 in 2020, in thousands.

Part 1 Guiding Questions:

(a) What is the annual change and annual percentage change of the population for the years 2015–2020?
(b) Create a scatter plot for the US population from 1950 to 2020. Add a linear trendline and display the trendline equation (keep 3 decimal places). Insert your graph here and write down your trendline equation. This is your **model**.
(c) You should observe that from 1950 to 2020, population growth in the USA has been nearly linear, indicated by your graph. Use your trendline equation to find the model population for the years 2000, 2010, and 2020.
(d) How good is the model? What is the error between actual population and model population for 2000, 2010, and 2020, respectively? What is the percentage error?

$$\textbf{Error} = \text{model} - \text{actual}$$

$$\textbf{Percentage error} = (\text{model} - \text{actual})/\text{actual}$$

(e) From your model, when will the population of the USA first reach 400 million people?
(f) Use your model to make population projections for the years 2050 and 2100.

Part 2. Ecological Footprint and Biocapacity

Recall that the ecological footprint represents the demand on our ecological resources, whereas the biocapacity represents the supply available. They are measured in the unit of "global hectare" (gha). A **global hectare** is a biologically productive hectare with world average biological productivity for a given year.

The Global Footprint Network tracks the ecological footprint and biocapacity of all countries since 1961 [24], using the United Nations data. The trends in the USA between 1961 and 2017 are plotted in Fig. 4, for the total and per capita gha, respectively. The USA is in a state referred to as overshoot.

Overshoot occurs when humanity's demand on the ecosystem exceeds its supply or regenerative capacity. This leads to a depletion of Earth's life supporting

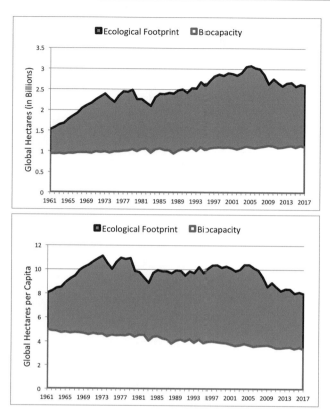

Fig. 4 The ecological footprint and biocapacity of the USA from 1961 to 2017. Top: total gha in billions and bottom: gha per capita. Data source: The Global Footprint Network 2021 [24]

natural resources and a buildup of waste. A country experiencing overshoot may have to import goods, energy, or other products.

Part 2 Guiding Questions:

(a) The total biocapacity of the USA is seen increasing slightly (Fig. 4 top), but the per capita biocapacity is declining (Fig. 4 bottom). Why?
(b) In 2017, the US per capita ecological footprint was approximately 8.0 gha per person, while the per capita biocapacity was approximately 3.4 gha per person. What was the biocapacity deficit? Include units.

Biocapacity Deficit (Reserve) = per capita biocapacity – per capita ecological

footprint, if negative (positive)

(c) A linear decreasing trendline is observed for the per capita biocapacity (Fig. 4 bottom). If this trend continues, what would happen?

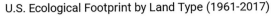

U.S. Ecological Footprint by Land Type (1961-2017)

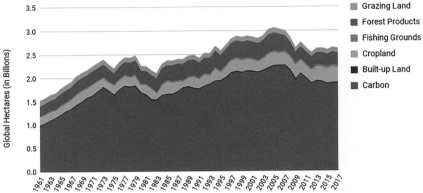

World Ecological Footprint by Land Type (1961-2017)

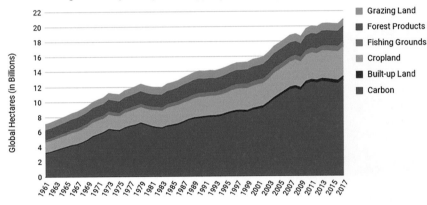

Fig. 5 US and world ecological footprint by land type from 1961 to 2017. Data source: The Global Footprint Network 2021 [24]

(d) Visit the Global Footprint Network Map [24]. Find another country with Biocapacity Deficit (−) and one country with Biocapacity Reserve (+) for 2017 (or the latest year available). Include units.

Country	Biocapacity (per capita)	Ecological footprint (per capita)	Biocapacity deficit (−) Biocapacity reserve (+)

Part 3. Carbon Footprint Reduction

The carbon footprint accounts for the majority of the ecological footprint in the USA, being over 70% in recent decades, as shown in Fig. 5, along with the world view. When the carbon footprint is reported within the context of total ecological footprint, the tonnes of carbon dioxide emissions are expressed as the amount of productive land area required to sequester those carbon dioxide emissions [23]. This tells us how much biocapacity is necessary to neutralize the emissions from burning fossil fuels. In 2017, the US per capita ecological footprint of consumption of carbon uptake land was 5.74 gha per person, while the world average was 1.69 gha per person [24].

How much carbon footprint in the USA needs to be reduced per decade with the objective of ending overshoot by 2050? To approach this question, let us hypothesize a policy scenario. For easier calculations, we will use the 2010 data instead of the most recent 2017 data. The total carbon footprint in 2010 was 2.06 billion gha.

Part 3 Guiding Questions:

(a) Suppose policies had been adopted after 2010 that would require a reduction of total carbon footprint by 18% every 10 years up to 2050. Would the total carbon footprint decrease linearly or exponentially over time?

(b) What would be your projections for the total and per capita carbon footprint for 2050? You will need to use the 2050 population projection obtained in Part 1 for the per capita estimation. Keep 2 decimal places.

(c) Assume that the per capita footprints for the rest land types are unchanged from 2010, the sum being 2.20 gha per person. What would the per capita ecological footprint be in 2050?

(d) The linear decreasing trendline (Fig. 4 bottom) predicts a per capita biocapacity of 2.52 gha per person in the USA in 2050. Compare this with your results from (c). Is overshoot being eliminated?

(e) More aggressive reductions must be required. What would you propose for the carbon footprint reduction, as well as other land types?

---

4   Climate Change

Carbon dioxide is the largest contributor of greenhouse gases, followed by methane. Human activities are emitting carbon dioxide into the atmosphere faster than natural processes can take it out. Over the past 170 years, the atmospheric concentration of CO_2 has been raised by 47% above the pre-industrial level found in 1850. The CO_2 levels today are higher than at any point in at least the past 800,000 years, as shown in Fig. 6. The ocean absorbs about 30% of the CO_2 that is released in the atmosphere. Ocean acidification is already harming many ocean species. The change is happening faster than at any time in geologic history.

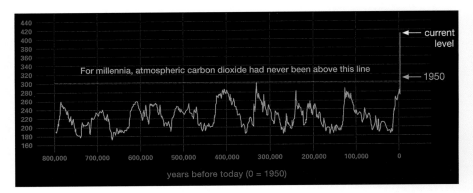

Fig. 6 Global atmospheric CO_2 concentrations (in ppm) for the past 800,000 years [36]

The Intergovernmental Panel on Climate Change (IPCC) was established in 1988 to provide policymakers with regular scientific assessments on climate change, its implications, and potential future risks, as well as to put forward adaptation and mitigation options [28]. The IPCC has made five assessment reports, with the fifth in 2014 [29]. The sixth report is to be released in 2022.

The Paris Climate Agreement, adopted by 196 countries on 12 December 2015, represented a huge historic step in reimagining a fossil-free future for our planet [50]. The treaty called for keeping global temperature rise well below 2 °C relative to pre-industrial levels by 2100, as well as agreeing to pursue efforts to limit the increase to 1.5 °C above pre-industrial levels. These bold moves suggest an end to fossil fuel use well before 2050.

An atmospheric concentration of greenhouse gases of 450 ppm carbon dioxide equivalent in 2100 gives us a 66% chance to comply with the 2-degree goal [29]. How close are we to reach the temperature limits set forth in the Paris Agreement? In 2020, we were already at 504 ppm carbon dioxide equivalent [40]. This sends an alarming message that we need to rapidly end emitting carbon.

How much difference will half-a-degree really make? The difference is huge. It means three times as many people exposed to extreme heatwaves and a massive 170% increase in flood risk compared to today, if a 2°C warming is reached [59]. How will the Arctic be affected? What does this mean for humanity and nature?

The year 2020 shattered numerous records for wildfires, heatwaves, hurricanes, droughts, and floods. The past six years were the hottest in the 170-year series. September Arctic sea ice is now declining at a rate of 13.1% per decade, relative to the 1981–2010 average [35]. The land ice sheets in Greenland have been losing mass since 2002, declining at a rate of 279.0 billion metric tons per year [37].

Think Can you come up with ways to put this huge number—279.0 billion metric tons—into perspective?

As a consequence of global warming, the added water from melting ice sheets and glaciers and expansion of sea water as it warms are the two primary factors causing sea level rise. Since 1880, sea level has risen over 9 inches (about 23 cm), with about three of those inches gained in the last 25 years [38].

Various possible future sea levels under several greenhouse gas emission and socioeconomic scenarios are projected by the IPCC, and, for the first time, the extensive data are made easily accessible to the public by NASA [39]. The high end of the emission spectrum yields projections with the most rapid rise in sea level. For example, the projected sea level rise under the "fossil-fueled development" scenario for Boston would reach 0.9 to 2.8 meters (about 3 to 9 feet) by 2150, using the 5th–95th percentile ranges. As sea level rises, many people, communities, and cities worldwide are being threatened. The dire impact of sea level rise is vividly illustrated in Fig. 7.

Research Project 2

Part 1. Rising of CO_2 Level and Global Temperature

Obtain the monthly mean concentrations of atmospheric carbon dioxide data measured at the Mauna Loa Observatory (1958 to present) from the Global Monitoring Laboratory (GML) of the National Oceanic and Atmospheric Administration (NOAA) [41]. Obtain the global land and ocean temperature anomalies time series (1880 to present) relative to the 1901–2000 global mean temperature from the National Centers for Environmental Information (NCEI) of NOAA [42].

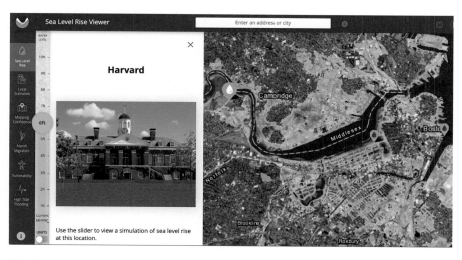

Fig. 7 Harvard (Cambridge, MA), the oldest university in the USA that has stood for almost four centuries, would stand in water in the next century if high emissions continued. Visualization: NOAA Sea Level Rise Viewer [43]

Part 1 Guiding Questions:

(a) What are the lowest and highest CO_2 concentrations? In which years did they occur?

(b) Plot the CO_2 dataset over time using a scatter plot. What is the annual rate of increase of the CO_2 concentration (the slope of the data)? At this rate, how long will it take for the CO_2 concentration to double from the present-day value?

(c) Plot the temperature anomalies dataset over time using a column chart. What are the five warmest years? How much warming has occurred? How much warmer will the Earth be in 100 years if this trend continues?

(d) Do you observe a relationship between the CO_2 level and temperature? Compare your data and plots over the same time period. Is there a strong correlation between the CO_2 level and temperature? You may use the CORREL function (Excel or Google Sheets) to compute the correlation coefficient between two variables.

> **Tip** A combo chart is a good option for displaying two chart types (such as column for temperature anomalies and line for CO_2) on the same chart.

> **Pearson's correlation** is a statistical measure that expresses the extent to which two variables are linearly related. The value ranges between -1 (strong negative relationship) and $+1$ (strong positive relationship). The values at or close to zero imply a weak or no linear relationship.

(e) You have examined data from the instrumented modern record. Investigate the records from earlier times in history. How do they compare with modern levels of CO_2 and temperature?

Part 2. On Thin Ice: Climate Change in the Arctic

Arctic sea ice reaches its minimum each September. On September 15, 2020, Arctic sea ice minimum set the second lowest on record, only after the record-low extent observed on September 17, 2012 [44].

While reading news one day you notice that two newspapers are running headlines on the decline of Arctic sea ice extent.

The Daily Alarm

"Arctic Sea Ice Extent Declines By Over 20%!"

The Chill Out Weekly

"Arctic Sea Ice Extent Decreases By 4% A Year"

Think Which of the two headlines do you think indicates a faster decrease of Arctic sea ice extent? Do they give you enough information to assess the rate at which the ice cap is disappearing?

Let us analyze the data and quantify change. Obtain the latest sea ice data from the National Snow & Ice Data Center (NSIDC) [44]. The Excel dataset (sea ice extent and area organized by year) contains monthly average Arctic sea ice extent (sheet 1) and area (sheet 2) for each year since 1979, in units of million square kilometers. We will use sheet 1 and sheet 2 only.

Part 2 Guiding Questions:

(a) For Arctic sea ice extent (sheet 1) and area (sheet 2), respectively,
   Compute the 1981–2010 average. This gives you the 30-year averaged data from January to December. In the same graph, plot this average (using scatter with smooth lines), along with data from the years 1979, 2012, 2014, 2019, and 2020. Add titles, axis labels, and legends. Include your graphs below. What do you observe?
   From now on, let us focus on the September Arctic sea ice extent (sheet 1).
(b) Plot the September data points from 1979 to 2020. Create a linear model by adding a linear trendline showing the equation. Add title and axis labels. Include your graph below. Write down the equation.
(c) What projection would you make for the future of Arctic ice cap? What does your model predict for the Arctic sea ice extent in the years 2030 and 2050? When would the Arctic sea ice disappear?
(d) What is the total change and percentage decrease of September Arctic sea ice extent from 2014 to 2020?
(e) What are the year-on-year percentage changes from 2014 to 2020? For example, 2014 to 2015, 2015 to 2016, and so on. Indicate increase or decrease. What is the average percentage change over this period?
(f) When did the most dramatic change occur during 2014 to 2020? How does this change compare to the average annual decrease?

(g) Revisit the two headlines. Were they wrong? Can you come up with a headline that describes the decline of Arctic sea ice extent since 2014 more accurately?

(h) Based on your data analysis, write a paragraph that gives a proper representation of the decline of Arctic sea ice extent since 2014.

5   Energy Production, Consumption, and Efficiency

The Lawrence Livermore National Laboratory (LLNL) has been maintaining the US energy flow charts since 2010 [33]. A single energy flow chart represents vast quantities of data, depicting resources, and their use. Energy resources include solar, nuclear, hydroelectric, wind, geothermal, natural gas, coal, biomass, and petroleum. The end-use sectors include residential, commercial, industrial, and transportation. Some of the basic source energy is first converted to electricity before it is transmitted to the end-use sectors. The energy flow diagram changes over time as new technologies are developed and as priorities change.

A sample flow chart is shown in Fig. 8. Of note is the huge amount of rejected energy, which is lost through waste heat. There is still much progress to be made in energy efficiency from the generating plant to the facility or home.

The energy measurements for the chart use units of **Quad**, which is quadrillion British thermal units (Btu). One quadrillion is 10^{15}. One Btu is the quantity of heat needed to raise the temperature of 1 lb of water by 1 °F at or near 39.2 °F. One barrel (42 gallons) of crude oil equals 5.7 million Btu, and one cubic feet of natural gas equals 1037 Btu [13].

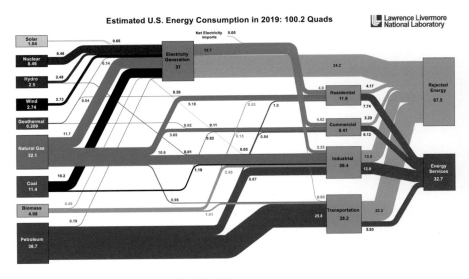

Fig. 8 2019 US energy flow chart (LLNL) [33]

Research Project 3

This project will guide you to examine the energy flows, productions, consumptions, efficiency, growth, and future of renewable energy sources.

Obtain the US energy flow charts for the year 2011 and 2019 from the LLNL [33]. The numbers on the chart are listed using three significant digits. Because of rounding, the total Quads listed for each production and end-use sector may not exactly equal the sum of the individual components.

Part 1. Energy Production

(a) Read the charts and rank the ten energy production sources from the highest to lowest in the following table. Create a bar chart to visualize the data. What do you observe?

Energy sources	2011	2019
1. Petroleum		
2. Natural gas		
3. Coal		
4. Nuclear		
5. Biomass		
6. Hydro		
7. Wind		
8. Solar		
9. Geothermal		
10. Electricity imports		
Total quads	97.3	100.2

(b) Petroleum ranks the highest in production. Petroleum is also called crude oil or just oil. It has mostly been recovered by oil drilling. How many quads of petroleum were supplied in 2011 and 2019, respectively? What is the growth (or decay) factor? What is the percentage change?

(c) The second highest is natural gas. Like petroleum, natural gas is obtained by drilling into underground hydrocarbon reservoirs. Natural gas is a hydrocarbon gas mixture consisting primarily of methane. What were the natural gas supplies in 2011 and 2019, respectively? What is the growth (or decay) factor? What is the percentage change?

(d) Which of the ten sectors of energy production are based on fossil fuels? Which represent renewable energy sources?

(e) What percentage of the total energy produced is due to renewables in 2011 and 2019, respectively? What is the growth rate of renewable energy?

Part 2. Energy Consumption and Efficiency

(a) Residential, commercial, industrial, and transportation are the four end-use sectors. The residential sector consumes energy through five different direct pathways. Rank them in the following table. Create a bar chart to visualize the data. What do you observe?

Energy forms	2011	2019
Electricity		
Natural gas		
Petroleum		
Biomass		
Geothermal		
Total quads	11.4	11.9

(b) The efficiency of an energy system is the percentage of the total energy used for the intended purpose. Determine the efficiency of electricity generation in 2011 and 2019, respectively.

(c) Can you think of a practical reason why so little electricity is distributed to the transportation sector?

(d) Most homes today are heated with electricity or natural gas. Electric heaters are 100% efficient since all of the energy that goes into the heater is turned into heat. Natural gas furnaces vary in efficiency, ranging from 78% to 98% [1], in turning the gas energy into heat, with 2%-22% wasted through furnace exhaust. Comparing these numbers, do you think that electric heaters are better than even the most efficient gas furnaces. Why or why not?

(e) What is the total energy rejected in 2011 and 2019, respectively? What is the total useful energy consumed? What is the efficiency of the entire US energy flow system?

(f) List one energy source that wastes (or rejects) the largest amount of energy—this is the most inefficient. List one energy source that wastes the least amount of energy—this is the most efficient. For the most inefficient, can you provide one change to make that energy source more efficient?

Outlook Will we reach the 50% reduction in CO_2 emissions by 2030 and close to net zero in 2050 required for a 1.5 °C warming future?

The energy sector is responsible for nearly three-quarters of the emissions that have already risen global temperature by 1.1°C above pre-industrial levels [14]. The energy sector has to be at the heart of the solution to climate change. If

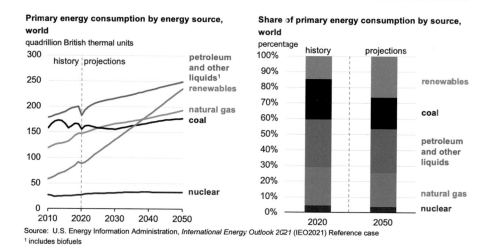

Source: U.S. Energy Information Administration, *International Energy Outlook 2021* (IEO2021) Reference case
[1] includes biofuels

Fig. 9 Projections of world primary energy consumption by source [14]. Despite a fast growth in renewables, liquid fuels remain the largest energy source

current policy trends continue, global energy consumption and energy-related CO_2 emissions will increase through 2050 as a result of population and economic growth. The international energy outlook by the US Energy Information Administration (EIA) projects a rapid growth in the renewable energy consumption, which more than doubles between 2020 and 2050 and nearly equals liquid fuels consumption by 2050, as shown in Fig. 9.

The world is electrifying. Electricity demand will be more than double over the next three decades, raising to 41% in the global energy mix in 2050. Electricity demand in the transportation sector will grow 26-fold from 2020 [8]. Projections of the world electricity generation by source are shown in Fig. 10. By 2050, more than 60% of the world's electricity will come from renewable sources. Solar photovoltaic (PV) will grow 12-fold and wind will grow fourfold from 2020; together they will contribute to over 40% of the electricity generated by 2050 (data source: EIA [14]).

CO_2 emissions from human activity peaked in 2019, which were brought forward five years by the impact of the COVID-19 pandemic in 2020. Despite a growing renewable share, the forecasted transition will not be fast enough for the world to achieve the ambitions of the Paris Agreement. We urgently need to find more sustainable and lasting ways to reduce CO_2 emissions.

6   Economic Growth (and Collapse)

In economics, **capital** refers to assets that are used to generate income, for example, a bridge that allows goods to flow to a new market, and a drug patent that allows a company to make money from a new treatment. The process by which new capital is created is called investment. The process by which capital wears out

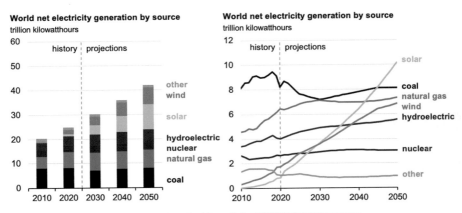

Fig. 10 Projections of world electricity generation by source [14]. Non-fossil sources will contribute to almost three-quarters of electricity generation by 2050

or becomes obsolete is called depreciation [47]. There are three types of capital: intellectual capital, social capital, and natural capital. Our natural capital is declining dramatically at the rates currently being exploited, such as the melting of glaciers and sea ice.

Economic prosperity is a very recent achievement of humanity. It can be measured by **GDP per capita**, which is the gross domestic product divided by midyear population. **GDP** is the sum of gross value added by all resident producers in the economy plus any product taxes and minus any subsidies not included in the value of the products [57]. Economic growth is the measure of the change of GDP.

In 1960, the global GDP per capita is estimated to be around $452 (current US$). In 2019, this value raised to $11,442 (current US$) [57]—more than 25 times that of 1960. This means that the output per person in one year in the past was about the output of an average person in two weeks today. The adjusted net national income per capita has grown from $695 (current US$) in 1970 to $9,290 (current US$) in 2018 [58]—the average person is 13 times richer than in 1970.

From Project 3, you have seen that the world population has increased threefold since 1950. Population growth and rising prosperity often go together in economic growth. Accompanying this economic growth is also the cost of environmental sustainability and social well-being. The GDP (or macroeconomic forecasts in general) ignores the degradation of natural capital and nature's constraints. Human actions are depleting Earth's natural capital, putting enormous strain on the environment. Humanity urgently needs to promote a sustainable pattern of economic and social development.

Research Project 4

In this project, you will investigate the relationships between GDP per capita and CO_2 emissions, as well as energy developments in relation to CO_2 emissions. We will revisit the concept of exponential growth and see how exponential processes interact with accumulations, which led to the great

Part 1. GDP per Capita, CO_2 Emissions, and Energy Consumption

Obtain the GDP per capita data (1960 to present) from the World Bank [57]. Obtain the energy and CO_2 emissions data (1965 to present) from the bp Statistical Review of World Energy [4].

(a) Plot world CO_2 emissions (y) vs. world GDP per capita (x) form 1965 to present (using a scatter plot). What do you observe?
(b) Create a linear regression (trendline). Compute the R-squared value (tutorial [20]) and covariance (tutorial [19]). Do the data reflect a strong correlation? You may display the R-squared value for the trendline.

> **R-squared** is a statistical measure of fit that indicates how much variation of a dependent variable is explained by the independent variables in a regression model.
>
> $$R^2 = 1 - \frac{\text{unexplained variation}}{\text{total variation}}$$
>
> **Covariance** evaluates how much two variables change together.

(c) Now, let us examine individual countries: China, the USA, and two countries of your choice. Plot the "primary energy consumption" and "primary energy consumption per capita" from 1965 to present. Some countries are small compared to China and the USA, so you may need more than one graph for the totals. One graph should work for per capita since it scales for population. What do you observe? What is the percentage increase for each country compared to 1965?
(d) Create a bar chart showing the fuel sources of total primary energy consumption of the most recent year for the two countries you selected, using sheet "Primary Energy—Cons by fuel." Which country has the highest reliance on renewables in absolute terms? And which in relative terms?
(e) For these two countries, respectively, plot CO_2 emissions (y) vs. total primary energy consumptions (x). Label each point with the year of observation. From the two graphs, which country appears to be better combatting CO_2 emissions? What factors might be contributing to success or failure?

Globally, primary energy consumption has been on the rise for at least half a century. Of note is that the primary energy consumption data include only commercially traded fuels (such as coal, gas, oil), nuclear and modern renewables (i.e., biofuels, geothermal, hydro, solar, and wind) and do not include traditional

Energy use per person, 2019

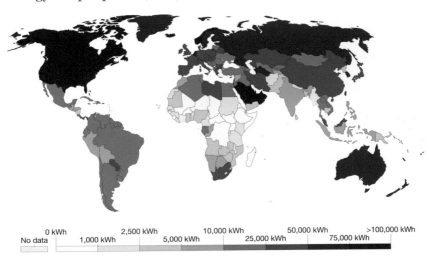

Source: Our World in Data based on BP & Shift Data Portal   OurWorldInData.org/energy · CC BY
Note: Energy refers to primary energy – the energy input before the transformation to forms of energy for end-use (such as electricity or petrol for transport).

Fig. 11 The per capita energy consumption across the world (2019), adapted from [45]

biomass (crop residues, wood, and other organic matter), which is often relied more on by poorer nations. The global view of per capita energy consumption is shown in Fig. 11. We see vast differences across the world. The largest energy consumers include Iceland, Norway, Canada, the USA, and wealthy nations in the Middle East such as Oman, Saudi Arabia, and Qatar. The average person in these countries consumes as much as 100 times more than the average person in some of the poorest countries [45].

Part 2. Case Study: Britain's Coal Reserves

In 1865, economist William Stanley Jevons examined Britain's coal supplies [47]. He estimated that Britain's mineable coal reserves were roughly 10^{11} tons and annual consumption was about 10^8 tons. If consumed at this constant rate, the reserves would last a thousand years. But Jevons noticed that annual consumption was not constant; in fact, it was increasing at a rate of approximately 3.5% per year—by now you should have recognized that this is an exponential increase. How long could Britain's reserves last if this exponential growth continued?

(a) The growth factor is $1 + 0.035 = 1.035$. The coal consumption for each successive year is thus 1.035 times that for the year before. Starting with 10^8 tons in 1865, what would the consumptions be in 1866, 1867, 1868, and in n years?

(b) What is the total coal consumption after n years? Write down the summation. Such a sum is called a **geometric progression**.

Geometric Progression Formula

$$1 + x + x^2 + \cdots + x^{n-1} = \frac{x^n - 1}{x - 1}, \quad x > 0 \text{ and } x \neq 1$$

(c) Compute the total coal consumption after n years using the formula.

(d) When will this total exceed 10^{11} tons? Set up an equation and solve for the value of n. You will need to use a logarithm.

Jevons estimated that if the growth continued, Britain's "thousand-year" coal reserves would be exhausted in about a hundred years. He wrote *"the check to our progress must become perceptible within a century from the present time; ... the conclusion is inevitable, that our present happy progressive condition is a thing of limited duration"* [30].

Jevons's work on coal reserves is one of the first known examples of calculation of an exponential reserve index. The **static reserve index** of a nonrenewable resource is the length of time it will last at the present rate of consumption. This indicates a linear growth. The **exponential reserve index** of a nonrenewable resource is the length of time it will last if consumption increases from year to year at a fixed exponential rate [47].

Part 3. Case Study: The 2008 Economic Collapse

Your generation escaped the direct impacts of the economic meltdown of 2008. Even decades later, the negative effects are still lingering. We will introduce the **Ponzi Scheme**, named after Charles Ponzi (1882–1949), that contributed to the downturn in the economy. Born and raised in Italy, Ponzi became notorious in the early 1920s as a swindler in North America for his money-making scheme, in which he offered his clients a 50% profit (or return) on certain "investments" within 45 days, or 100% within 90 days [56]. Let us understand the mathematical structure behind it.

(a) Ponzi had an investment opportunity and recruited 10 people to invest with him $1 each, which he guaranteed would become $1.50 in short order. How much money does Ponzi have?

(b) Before Ponzi could repay each dollar with a dollar and a half, he recruited 100 new investors to send him $1 each, so they could take advantage of this great deal. How much money does he have now?

(c) Ponzi had promised to send each of the 10 original investors $1.50. By doing so, Ponzi would create 10 success stories that could be used to motivate new investors. How much money does he have now? Let us call this "level 1."

(d) Ponzi then put out a call for 1000 new investors, while widely advertising the success of the original 10. The transactions kept going. If Ponzi honored the deal with both the first round of 10 and second round of 100 investors, what would be his net take? Let us call this "level 2."

(e) The next stage required 10,000 investors, but Ponzi now has 110 successful investors to advertise. Assume the recruiting and repaying went smoothly. What would be Ponzi's net take? Let us call this "level 3."

(f) This scheme continued to proceed. What pattern do you observe? Just about this time, the investors realized that at some point the miracle of making 50% profit would come to an end (where no new investors could be found to cover the scheme), Ponzi could run away with billions of dollars.

(g) Calculate the total amount of money taken in by Ponzi after reaching "level n." [Hint: it is the difference between two geometric progressions.] How large can this number become?

The mathematics allows us to make a powerful observation. The principal characteristic of the scheme is that via mathematical trickery, Ponzi rapidly accumulated large sums of money while doing nothing economically productive. Ponzi schemes are fraud and illegal.

In the real world, the ideal Ponzi pyramid will have variations, but regardless of the form, every Ponzi scheme exhibits growth of prodigious proportions. American financier and one-time chair of the NASDAQ stock exchange, Bernard Madoff, was able to run a Ponzi scheme for 15 or more years without interruption. He was eventually caught and sentenced to 150 years in prison in June 2009. Upon his arrest, $65 billion dollars was missing in his clients' accounts, and at least $18 billion of investors' initial capital was missing [56].

7   Policy and Social Justice

Economic growth brings a mix of benefits and costs to sociocultural well-being. Greenhouse gas emissions tend to increase with the level of economic activities. The Stern Review on the Economics of Climate Change (a 700-page report commissioned by the British government in 2005) indicates that stabilizing atmospheric concentrations of greenhouse gases at between 500 and 550 ppm of carbon-dioxide-equivalent would cost the equivalent of about 1% of global GDP per year [48]. 10 years later, Stern [49] finds a stronger case for a low-carbon economy and calls for a more demanding greenhouse gas target of no more than 500 ppm, in order to limit global average temperature increase to 2 °C. Achieving this target would require a gradual yet nearly complete transition away from today's high-carbon energy economy over the course of next several decades. A 1% reduction in economic output would result in an annual cost of roughly $200 billion in the context of current US economy, which is a substantial impact. Stern states [49]:

"The costs of inaction are much greater than the costs of action, on any sensible examination. And the 'greatest market failure the world has seen' does require strong policy."

The technological advances in renewables are remarkable. Electricity powered by renewables is the main driver of accelerating efficiency gains in the global energy system that will outpace both population and GDP growth [8].

Although associated, the climate problems are not caused by the economic growth, but rather by the absence of effective public policies to regulate greenhouse gas emissions. In the long run, the growth of material production and consumption is limited by natural resource constraints. Achieving a sustainable future will require policies and institutions that maintain the economy within bounds set by nature.

In today's world, there is a vast inequality between and within countries and societies, such as education, health, wealth, racial and gender differences. The **Human Development Index** (HDI) was created to emphasize that people and their capabilities should be the ultimate criteria for assessing the development of a country, not economic growth alone [53]. The **HDI** is a composite index on a scale from 0 to 1.0 that measures the average achievement in three key dimensions of human development: a long and healthy life, being knowledgeable, and have a decent standard of living. The HDI can be used to question national policy choices, asking how two countries with the same level of gross national income per capita can end up with different human development outcomes.

One powerful presentation is shown in Fig. 12 for the HDI vs. GDP per capita by country. An HDI value above 0.8 is classified as very high, between 0.7 and 0.799 as high, 0.55 to 0.699 as medium, and below 0.55 as low. The size of the dot is proportional to the population size. The purple dots at the lower left bottom, which are countries with higher poverty rates and gender inequality, are mostly in Africa. The best-off countries are mostly located in Europe and North America.

Figure 13 gives a global view of average years of schooling. The differences of education across the world are very large, ranging from the highest values in North America, Europe, Japan, and Oceania to the lowest in central Africa.

The most vulnerable nations to climate change are located primarily in Sub-Saharan Africa and South Asia and among the Small Island Developing States [12]. Fast-growing African cities are facing worse climate risks. They have the highest rates of urbanization and little to no environmental infrastructure and are set to pay the highest price of the consequence. The cities at extreme risk already have highly vulnerable populations and lack adequate healthcare services and disaster mitigation systems. The scale of the risk could threaten the capital flows that have streamed into these markets to take advantage of burgeoning economies, emerging consumers and cheap labor [55].

Research Project 5

In this final project, you will dive into social justice issues on the local or global scales. As indicated in Fig. 1, the three dimensions of sustainability—environmental protection, economic development, and social well-being—are inseparably linked. Social change and policy aimed at environmental stewardship and social equity is achieved through the interaction of behavioral change and structural change. In essence, the pathway to social justice and sustainability is to collectively confront the challenges and consequences of uneven development, inequality, and disparity, both in the political economic sense and in the environmental sense.

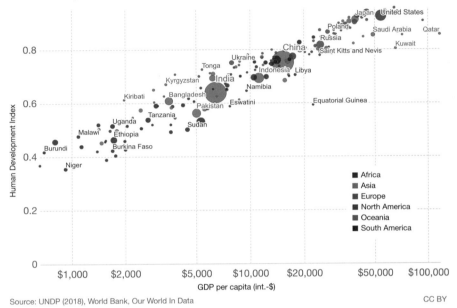

Fig. 12 HDI vs. GDP per capita (international-$) in log scale (2017), adapted from [46]

You will develop a project plan through investigation, design, action, reflection, and demonstration. You may gather data and create graphs, make analysis and findings, and use mathematical reasoning to learn and discover.

Think about how to ask refined quantitative questions. Choose appropriate mathematical ideas and tools to answer your questions. Develop conceptual frameworks to recognize, evaluate, and implement ways to challenge social sustainability. Be creative and open-minded. Think globally, act locally. Consider how you can best use your knowledge and voice to educate, inspire, and effect change.

Sample topics include:

- Educational access, funding, testing, achievement gap
- Food insecurity, hunger
- Gender discrimination, gender pay gap
- Health disparities, health care and insurance
- Housing, gentrification, homeownership
- Poverty, minimum/living wage, sweatshops
- Prisons, racial profiling, death penalty
- Privilege, power, social class
- Racism, segregation, stereotyping
- Welfare, civil rights, human rights

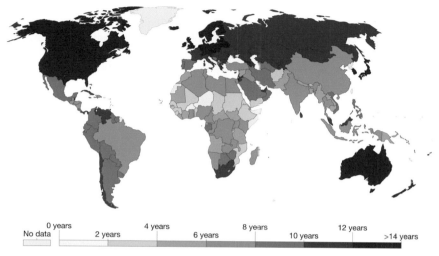

Fig. 13 Average years of schooling across the world (2017), adapted from [46]

Fig. 14 United Nations Sustainable Development Goals [52][1]

---

[1] Reprinted with permission of the United Nations. The content of this publication has not been approved by the United Nations and does not reflect the views of the United Nations or its officials or Member States.

We hope that through this project, you will increase global awareness and global perspective of the interrelationship of human activities and the Earth's natural systems, as well as your willingness to involve and engage in local, global, and intercultural sustainability problem solving.

In 2015, the United Nations set forth 17 global goals for a better world by 2030 [52]. They are known as the sustainable development goals, outlined in Fig. 14. These goals have the power to end poverty, fight inequality, and stop climate change. Guided by these goals, it is now up to all of us, governments, businesses, civil society, and the general public to work together to build a better future for everyone.

References

1. American Gas Association, Natural Gas Heating Systems. https://www.aga.org/natural-gas/in-your-home/heating
2. Berg, T.: Intermediate Algebra 1.6 Unit Conversion Word Problems. BC Open Textbook (2020). https://opentextbc.ca/intermediatealgebraberg/chapter/chapter-1-6-unit-conversion-word-problems
3. Boyle, K. and Örmeci, B.: Microplastics and Nanoplastics in the Freshwater and Terrestrial Environment: A Review. Water 12(9), 2633 (2020).
4. bp Statistical Review of World Energy 2021. https://www.bp.com/en/global/corporate/energy-economics/statistical-review-of-world-energy.html
5. Breeze, P. (editor): Power Generation Technologies (Third Edition). Newnes (2019).
6. Brundtland Report: Our Common Future. United Nations World Commission on Environment and Development (WCED). Oxford University Press, Oxford (1987). http://www.un-documents.net/wced-ocf.htm
7. Clause, A.G., Celestian, A.J. and Pauly, G.B.: Plastic Ingestion by Freshwater Turtles: A Review and Call to Action. Sci Rep 11, 5672 (2021). https://doi.org/10.1038/s41598-021-84846-x
8. DNV GL Energy Transition Outlook 2020. https://eto.dnv.com/2020
9. The Earth Charter Initiative (2000). https://earthcharter.org/read-the-earth-charter
10. European Chemical Transport Association (ECTA): Guidelines for Measuring and Managing CO_2 Emission from Freight Transport Operations. Issue 1 (2011).
11. Environmental Defense Fund (EDF+Business): The Green Freight Handbook (2019). https://supplychain.edf.org/resources/the-green-freight-handbook
12. Edmonds, H.K., Lovell, J.E. and Lovell, C.A.K.: A New Composite Climate Change Vulnerability Index. Ecological Indicators, 117:106529 (2020). https://doi.org/10.1016/j.ecolind.2020.106529
13. United States Energy Information Administration (EIA): Monthly Energy Review (2020). https://www.eia.gov/totalenergy/data/monthly/pdf/mer.pdf
14. United States Energy Information Administration (EIA): International Energy Outlook (2021). https://www.eia.gov/outlooks/ieo
15. Expert Market Research: Global Paper Cups Market Report and Forecast (2020). https://www.expertmarketresearch.com/reports/paper-cups-market
16. United States Environmental Protection Agency (EPA): Carbon Footprint Calculator. https://www3.epa.gov/carbon-footprint-calculator
17. United States Environmental Protection Agency (EPA): Inventory of U.S. Greenhouse Gas Emissions and Sinks. https://www.epa.gov/ghgemissions/inventory-us-greenhouse-gas-emissions-and-sinks

18. United States Environmental Protection Agency (EPA): Fast Facts on Transportation Greenhouse Gas Emissions. https://www.epa.gov/greenvehicles/fast-facts-transportation-greenhouse-gas-emissions

19. Microsoft Excel Covariance Function. https://support microsoft.com/en-us/office/covariance-p-function-6f0e1e6d-956d-4e4b-9943-cfef0bf9edfc

20. Microsoft Excel R-squared Function. https://support.microsoft.com/en-us/office/rsq-function-d7161715-250d-4a01-b80d-a8364f2be08f

21. Microsoft Excel Scatter Charts and Line Charts. https://support.microsoft.com/en-us/topic/present-your-data-in-a-scatter-chart-or-a-line-chart-4570a80f-599a-4d6b-a155-104a9018b86e

22. The Global Footprint Network: Ecological Footprint. https://www.footprintnetwork.org/our-work/ecological-footprint

23. The Global Footprint Network: Carbon Footprint. https://www.footprintnetwork.org/our-work/climate-change

24. The Global Footprint Network: Map and Data. https://data.footprintnetwork.org

25. Global Carbon Project (2020). https://doi.org/10.18160/gcp-2020

26. Google Sheets: Create and Edit Charts. https://support.google.com/docs/topic/1361474?hl=en

27. The Intergovernmental Science-Policy Platform on Biodiversity and Ecosystem Services (IPBES): Global Assessment Report on Biodiversity and Ecosystem Services. Brondizio, E.S., Settele, J., Daz, S. and Ngo H.T. (editors) (2019). https://ipbes.net/global-assessment

28. The Intergovernmental Panel on Climate Change (IPCC). https://www.ipcc.ch

29. The Intergovernmental Panel on Climate Change (IPCC): The Fifth Assessment Report, AR5 Climate Change 2014: Impacts, Adaptation, and Vulnerability. https://www.ipcc.ch/report/ar5/wg2

30. Jevons, W.S.: The Coal Question: An Inquiry Concerning the Progress of the Nation, and the Probable Exhaustion of Our Coal-Mines. Classic Reprint Series. Forgotten Books (2017). Reprinted from 2nd edition, originally published in 1866.

31. Chris Jordan Photography: Paper Cups 60 × 96″ (2008). http://www.chrisjordan.com/gallery/rtn/#paper-cups

32. Chris Jordan Photography: Blue 60 × 70″ (2015). http://www.chrisjordan.com/gallery/rtn2/#water-bottles

33. Lawrence Livermore National Laboratory (LLNL): Energy Flow Charts. https://flowcharts.llnl.gov/commodities/energy

34. National Geographic: How the Plastic Bottle Went from Miracle Container to Hated Garbage (2019). https://www.nationalgeographic.com/environment/2019/08/plastic-bottles

35. NASA Global Climate Change: Vital Signs of the Planet, Arctic Sea Ice Minimum. https://climate.nasa.gov/vital-signs/arctic-sea-ice

36. NASA Global Climate Change: Vital Signs of the Planet, The Relentless Rise of Carbon Dioxide. https://climate.nasa.gov/evidence

37. NASA Global Climate Change: Vital Signs of the Planet, Ice Sheets. https://climate.nasa.gov/vital-signs/ice-sheets

38. NASA Global Climate Change: Vital Signs of the Planet, Sea Level. https://climate.nasa.gov/vital-signs/sea-level

39. NASA Sea Level Projection Tool (2021). https://sealevel.nasa.gov/data_tools/17

40. National Oceanic and Atmospheric Administration (NOAA): Annual Greenhouse Gas Index. https://www.esrl.noaa.gov/gmd/aggi/aggi.html

41. National Oceanic and Atmospheric Administration (NOAA): Global Monitoring Laboratory (GML), Monthly Mean Concentrations of Atmospheric Carbon Dioxide. https://www.esrl.noaa.gov/gmd/ccgg/trends/data.html

42. National Oceanic and Atmospheric Administration (NOAA): National Centers for Environmental Information (NCEI), Global Land and Ocean Temperature Anomalies. https://www.ncdc.noaa.gov/cag/global/time-series

43. National Oceanic and Atmospheric Administration (NOAA): Sea Level Rise Viewer. https://coast.noaa.gov/slr

44. National Snow & Ice Data Center (NSIDC). https://nsidc.org/arcticseaicenews/sea-ice-tools
45. Ritchie, H. and Roser, M.: Energy Production and Consumption. Published online at OurWorldInData.org (2020). https://ourworldindata.org/energy-production-consumption
46. Roser, M.: Human Development Index (HDI). Published online at OurWorldInData.org (2019). https://ourworldindata.org/human-development-index
47. Roe, J., deForest, R. and Jamshidi, S.: Mathematics for Sustainability. Springer (2018).
48. Stern, N.: Stern Review on The Economics of Climate Change. HM Treasury, London (2006).
49. Stern, N.: Economic Development, Climate and Values: Making Policy. Proc. R. Soc. B. 282:20150820 (2015). https://doi.org/10.1098/rspb.2015.0820
50. United Nations: Climate Change. The Paris Agreement (2015). https://unfccc.int/process-and-meetings/the-paris-agreement/the-paris-agreement
51. United Nations: Department of Economic and Social Affairs, Population Division. World Population Prospects. https://population.un.org/wpp/Download/Standard/Population
52. United Nations: Sustainable Development Goals. https://www.un.org/sustainabledevelopment
53. United Nations Development Programme (UNDP): Human Development Reports, Human Development Index (HDI). http://hdr.undp.org/en/content/human-development-index-hdi
54. United States Geological Survey: How Much Water is There on Earth? https://www.usgs.gov/media/images/all-earths-water-a-single-sphere
55. Verisk Maplecroft: 84% of world's fastest growing cities face 'extreme' climate change risks (2018). https://www.maplecroft.com/insights/analysis/84-of-worlds-fastest-growing-cities-face-extreme-climate-change-risks
56. Walter, M.: Mathematics for the Environment. Chapman and Hall/CRC (2011).
57. The World Bank: GDP per Capita. https://data.worldbank.org/indicator/NY.GDP.PCAP.CD
58. The World Bank: Adjusted Net National Income per Capita. https://data.worldbank.org/indicator/NY.ADJ.NNTY.PC.CD
59. World Wildlife Fund (WWF): Our Warming World (2018). https://www.wwf.org.uk/updates/our-warming-world-how-much-difference-will-half-degree-really-make
60. World Wildlife Fund (WWF): The Lifecycle of Plastics (2018). https://www.wwf.org.au/news/blogs/the-lifecycle-of-plastics#gs.pa4gu0

Mosaics and Virtual Knots

Sandy Ganzell

Abstract

This chapter begins with a brief introduction to the mathematical theory of knots. We then describe Gauss codes, which are a way of encoding a knot without drawing pictures. It turns out that even though every knot can be given a Gauss code, not every Gauss code corresponds to a knot. A Gauss code that isn't a knot gives us a new kind of object called a virtual knot.

Both real knots and virtual knots can be studied by placing special square tiles (which have pieces of knots drawn on them) on to a grid to create a drawing called a mosaic. We will look at a new way of creating mosaics for virtual knots and study some of their properties.

Suggested Prerequisites *None. Really. This section sometimes requires visual imagination and creative problem solving, but there are no formal mathematical prerequisites.*

1   Math and Knots

Knots are familiar to anyone who has tied shoelaces, braided hair, or wrestled with tangled earbuds. Physical properties of knots are important to sailors, anglers, and rock climbers. The branch of mathematics known as *Knot Theory* studies questions related to all these and has other surprising applications in chemistry, molecular biology, and quantum physics. But you don't have to know anything about any of

S. Ganzell (✉)
St. Mary's College of Maryland, St Marys City, MD, USA
e-mail: sganzell@smcm.edu

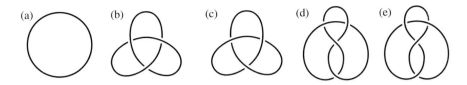

Fig. 1 Mathematical knots

those things to get started. First, we will explain what we mean by *mathematical knots*.

If you take a piece of rope and tie a knot in it, in principle (if you don't pull it too tight), you can always untie it. But if you glue the loose ends together, then probably you can't. When you glue the loose ends of a piece of rope together, you get a mathematical knot. We say two mathematical knots are the same (or equivalent) if you can manipulate one to look like the other without cutting and gluing.

Note: it's a good idea when you start studying knot theory to keep some rope or string handy, along with some tape for connecting the loose ends.

Exercise 1 Use string and tape to convince yourself that knots (a), (b), (c), and (d) in Fig. 1 are all different, but knots (d) and (e) are the same. Pay close attention to the pair (b) and (c), which really are different, and the pair (d) and (e), which really are the same.

Before going further, we will need a few comments about notation and terminology. When determining if two knots are the same or different, we ignore the length of the rope. So, a small, unknotted circle and a big, unknotted circle are considered equivalent. In Fig. 1, knot (a) is called the *unknot* or the *trivial knot*. Knots (b) and (c) are both called *trefoil* knots. Since they aren't equivalent, we sometimes call (b) the *left* trefoil and (c) the *right* trefoil. Knot (d), which is equivalent to (e), is called the *figure-8* knot.

Knots (b) and (c) are *mirror images* of each other. Since the trefoil is different from its mirror image, we say the trefoil is *chiral*. But the figure-8 is equivalent to its mirror image, so we call it *amphichiral*. Chirality is important in applications of knot theory to chemistry. Chemists often use the word *achiral* instead of amphichiral.

The drawings in Fig. 1 are called *knot diagrams*. The places where it looks like one portion of the knot is going under another part are called *crossings*. The diagram in Fig. 1b has three crossings, and diagram (d) has four. The smallest number of crossings necessary to draw a particular knot is called the *crossing number* of that knot. For example, the figure-8 can be drawn with 4 crossings but can't be drawn with fewer than 4, so its crossing number is 4. Similarly, the crossing number of each trefoil is 3. The unknot has crossing number 0.

Exercise 2 Draw a diagram of the trefoil that has four crossings. Draw diagrams of the unknot that have 1, 2, or 3 crossings.

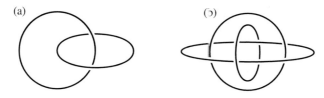

Fig. 2 A 2-component link (the Hopf link) and a 3-component link (the Borromean rings)

Exercise 3 Explain why there are no knots with crossing number 1. (Suggestion: draw a crossing, and then figure out how to connect the four loose ends without creating any new crossings. Which knot do you always get?) Try to use the same strategy to show that there are no nontrivial knots with exactly two crossings.

In the two-crossing part of Exercise 3, you may have found a diagram that looks like the one in Fig. 2a. That object cannot be made out of a single piece of rope, so it isn't considered a knot. Instead, we call it a *link*. Each closed loop of rope is called a *component* of the link. Sometimes, knots are considered links with only one component. In Fig. 2, the 2-component link (a) is called the *Hopf link*. The 3-component link (b) is called the *Borromean rings*.

Exercise 4 Convince yourself that the Borromean rings are in fact linked—you can't pull them apart.

Challenge Problem 1 An interesting property of the Borromean rings is that if you remove any one of the components, the other two will come apart to form an *unlink*. Try to find a 4-component link that has that property. In other words, find a 4-component link that really is linked, but if you remove *any one* of the components, the remaining three will come apart to form a 3-component unlink. Notice that you can't just add a fourth component to the Borromean rings because if you removed that fourth component, the remaining three will still be linked. If you can find a 4-component link, try to generalize so that you can do it with a link of any number of components.

To finish up this section, we will look at a way to combine two knots to get a new knot. The process is called the *connected sum*. If we draw two knot diagrams next to each other and then erase a small piece of each (not any of the crossings), we can connect the two knots with parallel strands to form a new knot as in Fig. 3.

The symbol we use for this operation is #. So, if we label the trefoil K_1 and the figure-8 knot K_2, then the connected sum would be written $K_1 \# K_2$. Knots that are formed by taking the connected sum of two nontrivial knots are called *composite* knots. Knots that aren't composite are called *prime*.

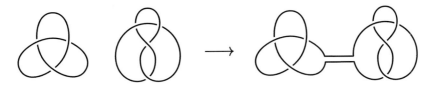

Fig. 3 Taking the connected sum of the right trefoil and the figure-8

Exercise 5 What do you get if you take the connected sum of the trefoil and the unknot?

There is a standard list of knots that mathematicians use for reference. It only has the prime knots, and if a knot is chiral, it only lists one of the mirror images. The list was mostly created by mathematicians Tait and Little in the 1890s, but it wasn't until the 1970s that it was rigorously shown to contain no duplicates and no omissions. Nowadays, it is usually called *Rolfsen's table of knots*, and it can't be found in many books and websites. Rolfsen's table lists all prime knots up to 10 crossings, and computer-assisted searches have catalogued all knots up to 16 crossings.

Rolfsen's table is organized by crossing number. So, the unknot comes first, then the trefoil and figure-8 knots, then the 5-crossing knots, and so on. For each crossing number, knots that have alternating diagrams (where the crossings alternate under and over as you follow the knot around) come before the non-alternating knots. (It turns out that some knots can't be drawn with an alternating diagram.)

The notation in Rolfsen's table is straightforward: 8_{19} refers to the 19th 8-crossing knot in the table. It just so happens that 8_{19} is the first non-alternating knot. The trefoil is 3_1, and the figure-8 is 4_1.

Exercise 6 Find a copy of Rolfsen's table. Draw diagrams for 5_1 and 5_2. You will get *a lot* better at drawing knots with some practice. Try 6_1, 6_2, and 6_3. Draw 8_{19} and notice that the diagram is non-alternating. It isn't easy to show, but it turns out that every possible diagram of 8_{19} is non-alternating.

2   Gauss Codes

Sometimes, it is useful to talk about which direction we are headed around a knot or a component of a link. We draw a small arrow on the knot (or several) to indicate the direction we are going, as in Fig. 4. A knot or link with a chosen direction is said to be *oriented*. Make sure when you draw your arrows they all point in a consistent direction around the knot.

Once a knot is oriented, each crossing will have two arrows pointing away: one on the strand that crosses over and one on the strand that crosses under. For each

Fig. 4 An oriented figure-8 knot and an oriented Hopf link

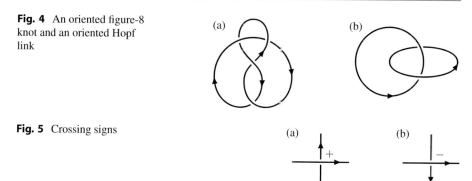

Fig. 5 Crossing signs

crossing, you can rotate your head (or if it's easier, rotate the paper) so that the over-strand points to the right, as in Fig. 5. The under-strand must then either point up or down. If it points up as in Fig. 5a, we say that the *sign* of the crossing is positive. If the under-strand points down as in Fig. 5b, we say that the sign is negative.

Exercise 7 Go back to Fig. 1. Choose orientations for the two trefoils and the figure-8 knot. Calculate the sign of each crossing. Now, reverse the orientation for each knot. (Change the direction of the arrows.) Did the crossing signs change or stay the same?

Exercise 8 Look at the oriented Hopf link in Fig. 4b. Find the crossing signs. Now, reverse one arrow but not the other. What happened to the crossing signs? What if you reverse both arrows?

It is often helpful to have a way to describe a knot without drawing a picture. For example, if you are using a computer to study knots, an input method that relies only on text can be advantageous. From a practical perspective, storing and searching through images of knots become increasingly challenging as the knots get more complicated. There are literally billions of different knots with 20 or fewer crossings.[1]

There are several ways of describing knots without drawing diagrams. The one we will describe here is called the *Gauss code*. Here is the procedure for finding a Gauss code for a knot. Note: this only works for knots, not for links. First, draw an oriented knot diagram, and label the crossings as in Fig. 6. It doesn't matter which crossing gets which label. You don't even have to use numbers for the labels, but most people do. Next, pick a starting point on the knot. In this example, we will use the dot at the very top. Now, follow the orientation around the knot until you

---

[1] Surprisingly, no one knows a formula for the exact number of knots with a given crossing number. But I wouldn't recommend spending too much time on that problem for now. It seems *very* hard.

Fig. 6 Calculating a Gauss
code for the figure-8 knot

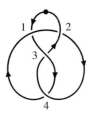

return to the starting point. Each time you pass a crossing, write down three things: whether you went over or under, the label, and the sign.

So, in our example, starting at the dot, we would go left (following the arrows) and go under crossing 1, which is a negative crossing. (Verify that crossing 1 is indeed negative.) So, the beginning of our Gauss code would be U1−. After that, we go over crossing 3, which is a positive crossing. So, we write O3+. Then, we go under crossing 4 (positive) and then over crossing 1, which we already determined is negative, then, under 2, over 4, under 3, and finally over 2. The complete Gauss code is U1−O3+U4+O1−U2−O4+U3+O2−.

Exercise 9 Find Gauss codes for the left and right trefoil knots (Fig. 1).

Note that if you choose a different labeling, a different starting point, a different orientation, or a different diagram, you will get different Gauss codes for the same knot.

Exercise 10 Describe how the Gauss code changes if you choose a different labeling. What if you choose a different starting point? What if you change the orientation?

Let's see how to reconstruct a knot from a Gauss code. We will use the example of the figure-8 knot above. The Gauss code begins U1−, so we will pick a starting point (marked with a dot) and draw an under-strand labeled 1.

Since crossing 1 is negative, the vertical segment (over-strand) must point upward. (Verify this.) Next comes O3+, so we continue from 1 and draw an over-strand labeled 3.

Crossing 3 is positive, so the vertical under-strand must point up. (Again, verify.) Continue the diagram with U4+.

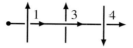

After that, we have to go over 1, so loop around and follow the arrow. Then, continue with U2−.

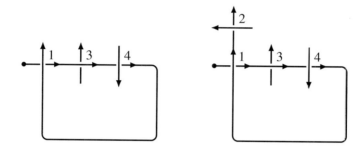

Finally, follow the arrows over 4, under 3, over 2. and back to the starting point.

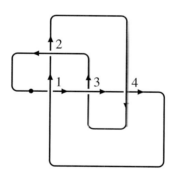

Exercise 11 Verify that the diagram above is equivalent to the figure-8 knot.

It's not easy to show, but it turns out that for a given Gauss code, there is only one knot with that code.

Exercise 12 Draw the knot with the following Gauss code:

$$U1-U3-O4-O1-U2-U4-O3-O2-$$

Which knot is it?

Let's point out two properties that Gauss codes must have.

1. Each label occurs exactly twice (once when we go over the crossing and once when we go under it).
2. For each label, the signs must be the same in both occurrences.

These two properties are not enough to guarantee that a given code is the Gauss code of a knot. Let us look at an example. We will attempt to draw a knot with Gauss code O1−O2+U1−U3−U2+O3−. Start with O1−O2+:

Then, we have to loop around to go under 1. As soon as we do that, we can see there is a problem.

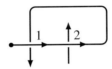

At some point, we are going to have to go under crossing 2, and then we will be trapped. We cannot add a crossing because there is nothing in the Gauss code between O2 and U1. In the next section, we will introduce a new kind of crossing that lets us escape from situations like this. The knot-like things we get will be called *virtual knots*.

There is another kind of diagram (called a Gauss diagram) that is sometimes helpful for studying knots when we have a Gauss code. To draw it, start by writing the labels of the Gauss code around a circle as in Fig. 7a. It doesn't matter where you start or which direction you go. In this example, we will use the Gauss code we saw earlier for the figure-8 knot: U1−O3+U4+O1−U2−O4+U3+O2−, starting near the top and going counterclockwise.

Fig. 7 Drawing a Gauss diagram for a knot

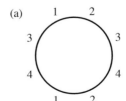

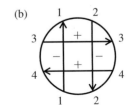

Fig. 8 Two equivalent Gauss diagrams (both for the figure-8 knot)

(a)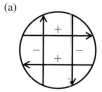

(b)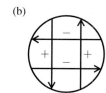

Fig. 9 Which knot is this?

Connect each pair of labels by drawing arrows *from the label representing the over-strand to the label representing the under-strand*, as in Fig. 7b. Finally, mark each arrow with a + or − to indicate the sign of the crossing. Remember that, in a Gauss diagram, the arrows represent the crossings of the knot—arrowheads are the under-strands and arrowtails are the over-strands. It doesn't matter precisely how you draw the arrows, as long as they have the correct starting and ending points.

Looking back at Exercise 10, if you choose a different labeling, it won't affect the arrows in the Gauss diagram, so we often just omit the labels. Choosing a different starting point for the Gauss code would just rotate the whole Gauss diagram. And reversing the orientation of the knot would just make our labels go clockwise instead of counterclockwise. Or you can just flip the Gauss diagram over like a pancake.

Exercise 13 The Gauss diagram in Fig. 8a is just the unlabeled Gauss diagram from the previous example. How can (b) be labeled to see that it is the same knot?

Exercise 14 Draw Gauss diagrams for each trefoil knot.

Exercise 15 Which knot is the Gauss diagram pictured in Fig. 9?

Challenge Problem 2 Explain why every arrow in a Gauss diagram must cross an even number of other arrows. Note that 0 is an even number, so the small arrow in the lower left portion of Fig. 9 is ok.

We will see in the next section that Challenge Problem 2 only applies to "real" knots—it is sometimes false for virtual knots.

3   Virtual Knots

Let's return to our attempt from the previous section to draw a knot with Gauss code $O1-O2-U1-U3-U2-O3-$. We got as far as $O1-O2-U1-$:

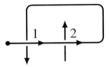

Let us continue and see what happens. We need to go under crossing 3, which has negative sign, and then under crossing 2.

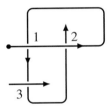

The Gauss code tells us to go over crossing 3 next, but there is no way to do that without adding another crossing. The way we get around this is by altering the paper: we cut out two circles and attach a hollow tube (called a *handle*) that gets us out of our trap. Then, it is easy to continue on to the O3 portion of the Gauss code.

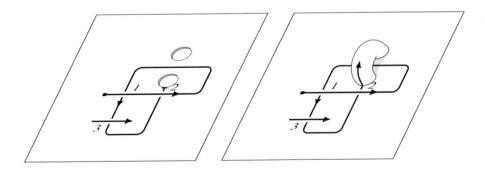

We escaped without really crossing the horizontal portion of the knot. It is important not to think of this as crossing over. In fact, you could have just as easily attached the handle underneath the paper. Instead, think of this as a new kind of crossing, neither over nor under. We call it a *virtual crossing*, and instead of attaching handles to our paper, we just indicate the crossings that are virtual with a small circle:

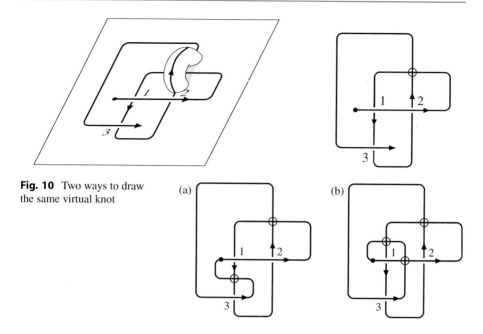

Fig. 10 Two ways to draw the same virtual knot

(a)

(b)

Remember that virtual crossings don't appear in the Gauss code. They don't get labels, and they don't have signs. It is like they aren't really there at all. To make a distinction, we sometimes call the "real" crossings *classical* crossings.

A subtle point here is that the paper we started with is a sort of three-dimensional—we draw some strands passing over and others passing under. But then the handles leave the dimension of the original paper. Here are two ways of thinking about this:

1. Think of the paper (including the paper that makes the handles we added) as having a little thickness—enough for one strand of the knot to pass over or under another. But the knot cannot leave the thickened paper. That is why the horizontal portion of the knot (O1−O2+) cannot be lifted up and over the virtual crossing. It would require pulling the knot out of the paper to get to the other side of the handle.
2. Think of virtual knots only as diagrams, *not* as objects that exist in three-dimensional space. When deciding whether two diagrams represent the same virtual knot, remember that the only changes you can make are (a) moves that do not affect the Gauss code and (b) moves that could be made for ordinary knots and do not involve the virtual crossings at all.

We will look more at these ideas after we finish the Gauss code from our example.

We have to get back to the starting point. To do that, we will need to add *another* virtual crossing, as in Fig. 10a. Or if we take a different route, we can add *two* virtual crossings as in Fig. 10b.

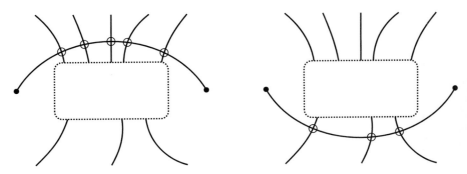

Fig. 11 The virtual detour move does not affect the Gauss code, so the two virtual knots are equivalent. It does not matter what is inside the box

Fig. 12 These moves are forbidden. No, really. They're called forbidden moves

To emphasize an important point: if two virtual knots have the same Gauss code, then they are equivalent. This lets us manipulate virtual knot diagrams in a variety of ways. See Fig. 11 for an example. Imagine that the box has some tangled mess of virtual and/or classical crossings. If a strand of a virtual knot has *only* virtual crossings (like the piece connecting the two dots), we can move it without affecting the Gauss code if the new crossings are all virtual also. This is called a *virtual detour move*. You can visualize the handle that connects the two dots—it can go above the box or below it. But it does not matter what is inside the box since that part of the virtual knot does not change when you do the virtual detour move.

There is one kind of move you have to avoid. You cannot move a strand with classical crossings across a virtual crossing as in Fig. 12. These moves are actually called *forbidden moves*. It turns out that if you allow the forbidden moves, and then all virtual knots are equivalent to the unknot.

Exercise 16 Draw a few examples of virtual knots. What is the effect of each forbidden move on the Gauss code? Draw the corresponding Gauss diagrams.

Exercise 17 Think about what forbidden moves would do to a virtual knot drawn on paper with attached handles.

Exercise 18 The virtual knot in Fig. 13 isn't the unknot. But show that you can unknot it if you use a forbidden move.

One strange note about terminology: knots that can be drawn with no virtual crossings are called *classical knots*. The term *virtual knot* is used for any Gauss code, whether the knot is classical or non-classical. In other words, every classical

Fig. 13 This virtual knot is
nontrivial. But it can be
unknotted using a forbidden
move

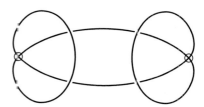

knot is virtual, but not every virtual knot is classical. A 3-crossing virtual knot has a
Gauss code with three classical crossings. There may be 0 or more virtual crossings.

Also, just like there is a standard table for classical knots, there is a table for
virtual knots. It's called Jeremy Green's table of Virtual Knots [5]. But there is
a catch. There are *a lot* of virtual knots. For classical knots, there is only one 3-
crossing knot (not counting mirror images) and only one 4-crossing knot. But there
are seven 3-crossing virtual knots and 108 different virtual knots with 4 crossings.
There are 2448 virtual knots with five crossings, but only two of them are classical.

Like Rolfsen's table, the numbering system for virtual knots is also somewhat
arbitrary. (There is actually a good reason for the way they are organized, but it
is not especially illuminating for our purposes.) Virtual knots are named by their
crossing number (remember only count classical crossings) and by their position on
the list. So, 3.1 is the first 3-crossing knot, and 4.17 is the seventeenth 4-crossing
knot on the list. The classical trefoil happens to be the sixth 3-crossing virtual knot,
so 3.6 is the same knot as 3_1. Yes—it's a bit unfortunate.

4   Mosaics

A *knot mosaic* is a square grid filled with any combination of the following eleven
tiles that forms a knot (or link) diagram:

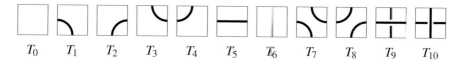

T_0 T_1 T_2 T_3 T_4 T_5 T_6 T_7 T_8 T_9 T_{10}

If you prefer, you can say there are just 5 tiles (including the blank tile) and you
are allowed to rotate them. But in this chapter, we will refer to the eleven tiles as
numbered above. Some pictures will make this clear. Figure 14 illustrates mosaics
for the figure-8 knot and knot 7_2.

Note that in order for the mosaic to form a knot (or link) diagram, adjacent tiles
must have their loose ends match up. (We sometimes say the mosaic is *suitably
connected* if there are no loose strands.) So, for example, tiles T_5 and T_6 cannot be
adjacent. The crossing tiles T_9 and T_{10} cannot be on an edge of the mosaic. The
top-left corner can only have tile T_0 or T_2.

Fig. 14 Knot mosaics for
knots 4_1 and 7_2

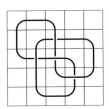 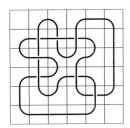

Exercise 19 Show that the trefoil can be realized (drawn) on 4×4 mosaic but not on a 3×3 mosaic. Repeat the exercise for the Hopf link.

Exercise 20 Explain why the figure-8 knot cannot be realized on a 4×4 mosaic. Suggestion: since the crossing number for the figure-8 is 4, there must be at least 4 crossing tiles (T_9 and T_{10}). Where do they have to go? What happens next?

If a knot can be realized as an $n \times n$ mosaic, but not on any smaller square, we say the *mosaic number* of the knot is n. So, considering Exercises 19 and 20, the mosaic number for the trefoil is 4, and the mosaic number for the figure-8 is 5. The mosaic number has been calculated for all knots with 8 crossings or fewer. See [9] for details.

Research Project 1 Calculate the mosaic numbers for all 9-crossing knots. Calculate the mosaic numbers for all 10-crossing knots. Some computer assistance may be needed.

Research Project 2 The mosaic number for the trefoil is 4, and the mosaic number for the connected sum of two trefoils is 5 (verify this). Can you find a formula for the mosaic number of a connected sum of n trefoils? What about a connected sum of n figure-8 knots?

If you know that the mosaic number of K_1 is m, and the mosaic number of K_2 is n, can you find a formula for the mosaic number of $K_1 \# K_2$? If not an explicit formula, can you find an inequality? For example, it is not hard to show that the mosaic number of $K_1 \# K_2$ is less than or equal to $m + n$. How much can you improve that?

The knot 6_1 is an interesting example. Its mosaic number is 5, but strangely, there is no 5×5 mosaic of 6_1 with only six crossings. To get the smallest mosaic, you need a diagram with seven crossings. See Fig. 15.

Fig. 15 Knot 6_1 has mosaic number 5, but we need to use a diagram with seven crossings

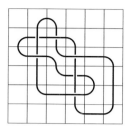 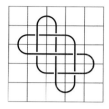

Exercise 21 Verify that the two knots in Fig. 15 are both 6_1.

There are other knots with this property: the smallest mosaic uses a diagram where the number of crossings is not minimal. But in all known examples, if the mosaic number of the knot is m, then the knot can fit on an $(m + 1) \times (m + 1)$ mosaic with its minimal number of crossings. For example, knot 6_1 has mosaic number 5, and a 6-crossing diagram for 6_1 fits on a 6×6 mosaic.

Research Project 3 Can you find a knot whose crossing number is c and whose mosaic number is m, but the smallest mosaic for the knot with exactly c crossing tiles is $(m + 2) \times (m + 2)$? Or to say that another way, can you find a c-crossing knot where you can reduce the minimal mosaic size by 2 (or more) by using a diagram with more than c crossings?

We do not have to limit ourselves to square mosaics. Instead of considering only $n \times n$ mosaics, let us allow $m \times n$ mosaics (where m and n may be different). Here are a few questions about rectangular mosaics.

Exercise 22 We have already seen that the trefoil fits in a 4×4 mosaic. Can the trefoil fit in a 3×5 mosaic? Such a mosaic would have smaller area than the 4×4 example.

Research Project 4 Using rectangular mosaics (not necessarily square), try to minimize the area of a mosaic. Figure 16 has an example of a knot with mosaic number 5 (area 25), which can be realized on a 4×6 mosaic (area 24).

- For each knot in Rolfsen's table, find the minimum area of a rectangular mosaic.

(continued)

Fig. 16 The connected sum of two trefoils has mosaic number 5 but can be realized on a mosaic with only 4 rows

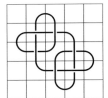
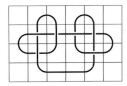

- Which knots have their minimum area realized by a square mosaic?
- Can you find an infinite set of knots whose minimal-area rectangular mosaics are squares?
- Can you find an infinite set of knots whose minimal-area rectangular mosaics are *not* squares?

Exercise 23 Which knots can be realized on a mosaic with only 3 rows? (Any number of columns is allowed.)

Research Project 5 Let us call the minimum number of rows required to make a rectangular mosaic of a knot the *minimal mosaic height* of the knot. So, for example, in Exercise 23, we saw that we cannot make a trefoil mosaic with only 3 rows. But we can make one with 4 rows. So, the minimal mosaic height of the trefoil is 4. Figure 16 has another example.

- Which other knots have minimal mosaic height 4? Which knots have minimal mosaic height 5?
- Find the minimal mosaic height for each knot in Rolfsen's table.
- Is there some number n where *every* knot has minimal mosaic height n or less? Or can you find a set of knots whose minimal mosaic height keeps getting bigger?
- For the trefoil, the mosaic number is 4 and the minimal mosaic height is also 4. Which other knots have mosaic number equal to their minimal mosaic height? Which knots have mosaic number exactly one more than their minimal mosaic height? Exactly 2 more?

5   Virtual Mosaics

Time to put it all together. The goal of this section will be to understand *virtual* mosaics. One way to do this would be to introduce a new tile with a virtual crossing:

This has been studied before. (See Sect. 6.) But there is another way to create a virtual mosaic that does not use any new tiles. Since virtual crossings are not "really" there, they will not really be in the mosaics either. Instead, we will connect parts of the mosaic to other parts like when we added handles. We will always connect each edge of the mosaic to exactly one other edge of the mosaic and always on the border. That way all the classical crossings will appear inside the mosaic, but the virtual crossings won't. See Fig. 17.

The labels on the border tell you how to connect the pieces of the mosaic to form a virtual knot diagram. All crossings outside the mosaic will be virtual. It doesn't matter how you draw the strands outside the mosaic since the virtual knots will all have the same Gauss code. So, they are the same knot. In fact, all the different ways of connecting the labels are related by virtual detour moves.

Here is an example of how to make a virtual mosaic from a Gauss code. We will make virtual knot 3.1, which has Gauss code $O1-O2-U1-O3+U2-U3+$. It is pretty easy to do on a large mosaic, but like the mosaics for classical knots, we are going to try to fit knot 3.1 onto as small a virtual mosaic as possible. Since there are three crossings, the smallest square we could hope for would be 2×2.

Let us try putting crossing 1 in the top-left pointing downward. Since it is a negative crossing, the horizontal portion must point to the left. (We will remove the arrows and numbers later, but they will help as we're building the mosaic.)

After that we have $O2-$, so we will put a T_{10} tile in the lower left and remind ourselves that the horizontal strand points left.

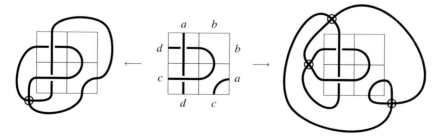

Fig. 17 Virtual knot diagrams can be created by connecting edges of a virtual mosaic. No matter how you connect the paired edges, you get the same virtual knot. Check the Gauss code if you are not sure

Next, we have to go under crossing 1. There are a few ways to do this, but we have to leave the mosaic and loop around to another edge piece like we did in Fig. 17. Since we need to follow the arrow at crossing 1, let us put a T_4 tile in the upper right corner and label the corresponding edges. Note: we could have tried a T_5 tile there and put the second "a" on the right edge (but see Exercise 24).

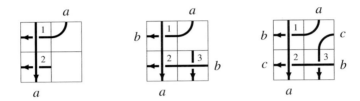

Looking at the next part of the Gauss code, after we go under crossing 1, we have O3+U2−, so we will need to leave the mosaic again. Label the edge with a "b" and reconnect at the lower right edge, crossing over 3. Since crossing 3 is positive, the vertical strand must point downward. Continue under crossing 2.

To finish the virtual mosaic, we need to go under crossing 3 next. So, label the "c" edges and change the T_4 tile to a T_8 tile (which doesn't affect the "a" edge). The completed virtual mosaic will include the "d" edges, and we remove the arrows and crossing numbers as in Fig. 18.

Exercise 24 Can a 2×2 virtual mosaic for knot 3.1 contain a T_5 tile?

Exercise 25 Find virtual mosaics for the rest of the 3-crossing virtual knots.

- 3.2: O1−O2+U1−O3−U2+U3−
- 3.3: O1−O2−U1−U3−U2−O3−
- 3.4: O1−O2+U1−U3−U2+O3−
- 3.5: O1−O2−O3−U1−U2−U3−
- 3.6: O1−U2−O3−U1−O2−U3−   (Note: this is the classical trefoil knot.)
- 3.7: O1−U2−O3+U1−O2−U3+

All of them can be completed on 2×2 mosaics.

Fig. 18 A completed virtual mosaic for virtual knot 3.1

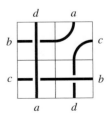

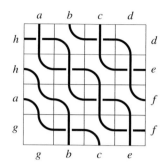 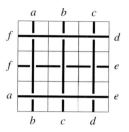

Fig. 19 Knot 7_1 as a 7-crossing 4×4 mosaic and as a 9-crossing 3×3 mosaic

Exercise 26 Find a 3×3 virtual mosaic for the classical figure-8 knot.

Challenge Problem 3 Show that there is no 2×2 virtual mosaic for the figure-8 knot.

Exercise 27 Since a 2×2 virtual mosaic only has four tiles, you can't make any knots with more than four crossings. But you can make all the classical 5-, 6-, and 7-crossing knots from Rolfsen's table on 3×3 virtual mosaics. Try 5_1, 5_2, 6_1, 6_2, and 6_3. Some of the 7-crossing knots are tricky. We will talk about those next.

Classical knot 7_1 is especially interesting. There is no 3×3 virtual mosaic for 7_1 with only seven crossing tiles. But there is a 3×3 virtual mosaic for 7_1 with *nine* crossings. See Fig. 19.

Exercise 28 Verify that both knots in Fig. 19 are knot 7_1. Notice that when you connect the corresponding edges, you can do it without creating any virtual crossings.

Knot 7_3 is similar. There is a 4×4 virtual mosaic with seven crossings, but there is a 3×3 virtual mosaic with eight crossings.

Challenge Problem 4 Find 3×3 virtual mosaics for knots 7_2 through 7_7.

Research Project 6 Knots 7_1 and 7_3 are examples of knots where the smallest virtual mosaic has more crossings than the crossing number of the knot. Knots 8_7, 8_{10}, and 8_{19} are also examples illustrating this phenomenon. Can you find an infinite set of such examples?

The smallest number n where there is an $n \times n$ virtual mosaic for the knot K is called the *virtual mosaic number* of K. So, for example, the virtual mosaic number of each 3-crossing virtual knot is 2, and the virtual mosaic number for each 7-crossing classical knot is 3.

A computer search of all the 3×3 virtual mosaics has shown that the only 8-crossing classical knots with virtual mosaic number 3 are 8_5, 8_7, 8_8, 8_{10}, 8_{12}, 8_{13}, 8_{14}, 8_{15}, 8_{19}, 8_{20}, and 8_{21}. All the other 8-crossing classical knots have been verified to have virtual mosaic number 4. For 9-crossing classical knots, only 9_{16}, 9_{23}, and 9_{31} have virtual mosaic number 3.

> **Research Project 7** Do all the other 9-crossings knots in Rolfsen's table have virtual mosaic number 4? What about all the 10-crossing knots?

If we have a classical mosaic for a knot, we can make a virtual mosaic as in the following example: Starting with a 6×6 mosaic for knot 7_2 (which happens to be the smallest for that knot), we can remove the top and bottom rows, along with the first and last columns, and then connect the loose ends in pairs, following the original mosaic as in Fig. 20.

So, if the mosaic number of a classical knot is n, then the virtual mosaic number is at most $n - 2$. But the virtual mosaic number could be even smaller. Even though the (classical) mosaic number for 7_2 is 6, the virtual mosaic number is actually 3, as shown in Fig. 21.

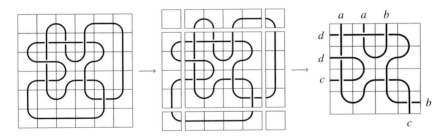

Fig. 20 Constructing a 4×4 virtual mosaic from a 6×6 classical mosaic

Fig. 21 Knot 7_2 has mosaic number 6, but its virtual mosaic number is 3

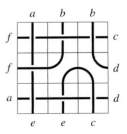

Research Project 8 Is there a knot with mosaic number n and virtual mosaic number $n - 4$? What about $n - 5$? Is there an example with virtual mosaic number $n - k$ no matter what k is?

Mosaics (classical and virtual) can be made for links as well as knots. Here is a virtual mosaic for the Hopf link:

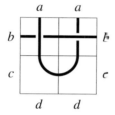

Research Project 9 Which links have virtual mosaic number 2? Which links have virtual mosaic number 3?

Look at the two virtual mosaics:

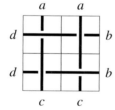 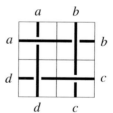

Both have the same *pattern*. In other words, the same tiles are used in the same places. But because the labels are in different places, they aren't the same knots. The first one is actually a link. But the second is the unknot.

Research Project 10 For a given pattern, what percentage of labelings produce the unknot? Which patterns have the highest percentage of unknots?

6   Further Reading

Perhaps the reason Knot Theory is such a good subject for undergraduate research is its accessibility. Beginning students can quickly understand open questions in the subject without sacrificing rigor. As the student learns more mathematics, new tools become available for studying the same questions, and more topics within Knot Theory become accessible. Precalculus is enough to study *polynomial invariants* of knots. After Calculus, students can study *hyperbolic knots*. Abstract Algebra leads to important questions about knots and groups; topology opens the doors to higher dimensional knot theory.

Two good introductory books on Knot Theory are *The Knot Book* by Colin Adams [1] and *An Interactive Introduction to Knot Theory* by Allison Henrich and Inga Johnson [6]. If you were able to understand this chapter, then you will be able to get through most of the chapters in those two books. Both have lots of accessible open questions. *The Knot Book* covers only classical knots but includes an introduction to hyperbolic knots and applications to Biology, Chemistry, and Physics. *An Interactive Introduction to Knot Theory* investigates both classical and virtual knots and introduces a variety of combinatorial ideas suitable for continuing research projects.

Knots and Links by Dale Rolfsen [13] is a classic textbook but is written for advanced undergraduates or beginning graduate students. At the very least, the student would need a course in Abstract Algebra before diving into it. But the hand-drawn illustrations are wonderful, and many knot theorists have a battered old copy on their shelf that (50 years later) still sheds new light on some old ideas. Rolfsen's table of knots is included as an appendix but is also available in many places online:

- http://katlas.org/wiki/The_Rolfsen_Knot_Table
- https://en.wikipedia.org/wiki/List_of_prime_knots

Jeremy Green's table of virtual knots is available online as well:

- https://www.math.toronto.edu/drorbn/Students/GreenJ/

Virtual Knot Theory was first presented by Lou Kauffman in the European Journal of Combinatorics [7]. Kauffman together with Sam Lomonaco developed the original idea of a knot mosaic in the article *Quantum knots and mosaics* [10]. More accessible papers on classical mosaics include *Knot mosaic tabulation* [9], which has significant contributions by undergraduates, *Mosaic Number of Knots* [8], and the paper *An infinite family of knots whose mosaic number is realized in non-reduced projections* [11], which solves the classical mosaic version of one of the research projects in this chapter.

Two articles looking at virtual mosaics with an additional "virtual crossing tile" are *Virtual Mosaic Knots* [4] and *Bounds on Mosaic Knots* [2]. More about the virtual mosaics in this chapter can be found in the paper *Virtual Mosaic Knot Theory* [3].

Finally, Kyle Miller [12] has developed some very nice online tools for identifying classical and virtual knots from drawings:

- https://kmill.github.io/knotfolio/
- http://tmp.esoteri.casa/virtual-knotfolio/

References

1. Adams, C. C. *The Knot Book, An Elementary Introduction to the Mathematical Theory of Knots*. (American Mathematical Society, 2004).
2. Alewine, A., Dye, H., Etheridge, D., Garduño, I. and Ramos, A. Bounds on Mosaic Knots. *crXiv:1004.2214* (2010).
3. Ganzell, S. and Henrich, A. Virtual Mosaic Knot Theory *J. Knot Theor. Ramif.*, 2020. https://doi.org/10.1142/S0218216520500911
4. Garduño, I. Virtual mosaic knots. *Rose-Hulman Undergrad. Math. J.* **10** no. 2 (2009): 5.
5. Green, J. A Table of Virtual Knots. https://www.math.toronto.edu/drorbn/Students/GreenJ/
6. Henrich, A. and Johnson, I. An interactive introduction to knot theory. Dover publications, Inc., 2017.
7. Kauffman, L. Virtual knot theory. *European J. Combin.* **20** no. 7 (1999): 663–690.
8. Lee, H. J., Hong, K., Lee, H. and Oh, S. Mosaic number of knots. *J. Knot Theor. Ramif.* **23** no. 13 (2014): 1450069.
9. Lee, H. J., Ludwig, L., Paat, J. and Peiffer, A. Knot mosaic tabulation. *Involve.* **11** no. 1 (2017): 13–26.
10. Lomonaco, S. J. and Kauffman, L. H. Quantum knots and mosaics. *Quantum Information Processing.* **7** no. 2–3 (2008): 85–115.
11. Ludwig, L., Evans, E. and Paat, J. An infinite family of knots whose mosaic number is realized in non-reduced projections. *J. Knot Theor. Ramif.* **22** no. 07 (2013): 1350036.
12. Miller, K. Virtual KnotFolio. http://tmp.esoteri.casa/virtual-knotfolio/
13. Rolfsen, D. Knots and Links. Mathematics Lecture Series, No. 7. Publish or Perish, Inc., Berkeley, Calif., 1976.

Graph Labelings: A Prime Area to Explore

Elizabeth A. Donovan and Lesley W. Wiglesworth

Abstract

A graph labeling assigns integers to the vertices or edges (or both) of a graph, subject to certain conditions. Labelings were first studied just over 50 years ago, making them a relatively new area of mathematics to explore. There are hundreds of different types of labelings, many of which are motivated by circuit design, coding theory, and communication networks. In this chapter, we introduce the reader to prime labelings, a labeling of the vertices in such a way that the labels of adjacent vertices are relatively prime. We also introduce several variations of prime labelings while guiding the reader through examples and existing research. We present the reader with several research questions for each variation, a few of which have never been explored, thus allowing the reader to make progress on new problems.

Suggested Prerequisites *Familiarity with set notation and a basic understanding of functions are sufficient to get the reader started. No additional prerequisites are required, though prior exposure to mathematical proofs and a basic understanding of modular arithmetic may be helpful.*

E. A. Donovan (✉)
Department of Mathematics and Statistics, Murray State University, Murray, KY, USA
e-mail: edonovan@murraystate.edu

L. W. Wiglesworth
Department of Mathematics, Centre College, Danville, KY, USA
e-mail: lesley.wiglesworth@centre.edu

© The Author(s), under exclusive license to Springer Nature Switzerland AG 2022
E. E. Goldwyn et al. (eds.), *Mathematics Research for the Beginning Student,*
Volume 1, Foundations for Undergraduate Research in Mathematics,
https://doi.org/10.1007/978-3-031-08560-4_4

1   Introduction

Graph theory is a relatively new branch of mathematics. It was first introduced in the 1700s by Leonhard Euler, whose solution to the famous Königsberg bridge problem is commonly considered the origin of graph theory. We present this problem as an introductory exercise below.

Exercise 1 The city of Königsberg (presently Kaliningrad, Russia) was situated on the Pregel River. The river runs through the town in such a way that two islands are formed. The four bodies of land are connected by a total of seven bridges as seen in Fig. 1. The question posed to Euler was whether it was possible to take a walk through the town in such a way that every bridge was crossed over exactly once and one could arrive back at the starting point?

Euler recognized that the relevant constraints were the four bodies of land and the seven bridges that connected them. In other words, the size of land and its location were not relevant. He then drew the first known visual representation of a modern graph, shown in Fig. 2. Each piece of land is represented by a point, called a *vertex* (plural: *vertices*), and each bridge is represented by a line, called an *edge*. The problem posed to Euler was thus equivalent to whether the graph below could be traced in such a way that every edge was traced over exactly once, starting and finishing at the same vertex. This "tracing" is commonly referred to as an Euler circuit.

Like Fig. 2, a graph is essentially a way of specifying relationships among a collection of items. The items can be represented as the vertices and the relationships as the edges. If an edge exists between two vertices, we say the vertices are *adjacent* or the vertices are *neighbors*.

Graph theory is used to model and study many things that affect our lives such as shipping and transportation routes, chemical molecular bonds, social networks, integrated circuits, and much, much more! To learn more about graph theory and its applications outside of the information given in this chapter, we recommend *The*

Fig. 1 A map of Königsberg from [12]. The seven bridges are highlighted in red

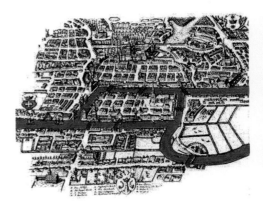

Fig. 2 A graphical
representation of Königsberg.
Note that the four vertices
represent the four landmasses
while the seven edges
represent the seven bridges

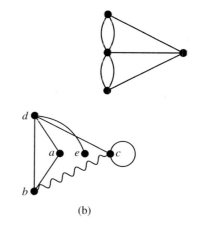

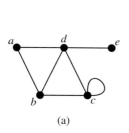

(a) (b)

Fig. 3 Two different visual representations of a graph G with vertex set $V(G) = \{a, b, c, d, e\}$
and edge set $E(G) = \{ab, ad, bc, bd, cc, cd, de\}$. Note that in (b) edges cd and de cross, but this
intersection does not form a vertex. Also since vertex c has a loop—an edge whose endpoints are
the same vertex—G is not a simple graph. (a) A representation of G. (b) A different representation
of G

Fascinating World of Graph Theory by Benjamin, Chartrand, and Zhang [4] as well
as Trudeau's *Introduction to Graph Theory* [21]. For a more comprehensive text,
you may consider West's *Introduction to Graph Theory* [24].

Unlike graphing a function, a graph G is a structure that does not only have one
visual representation. Since G really just consists of a set of objects (the vertices,
denoted $V(G)$) and a set of pairs (the edges, denoted $E(G)$), the only constraint
when drawing a graph is that the edge sets are the same. In Fig. 3, we can see two
different visualizations of the same graph as both graphs have the same vertex set
$\{a, b, c, d, e\}$ and the same edge set $\{ab, ad, bc, bd, cc, cd, de\}$. (Note that while
an edge is actually a set of two vertices, $\{v_1, v_2\}$, this notation has been shortened,
for readability, to $v_1 v_2$, for all $v_1 v_2 \in E(G)$.)

Notice the difference in the edges between the graph given in Figs. 2 and 3.
In Fig. 2, the graph contained *multiple edges*—more than one edge with the same
endpoints. In Fig. 3, there are no multiple edges (and no loops!) and so G is a *simple
graph*. We will only consider simple graphs—graphs without loops or multiple
edges—throughout the remainder of this chapter.

The number of edges *incident* to a vertex is the degree of that vertex. Notationally,
for any vertex $v \in V(G)$, the degree of v is denoted as $\deg(v)$. A vertex of degree
one is called a *pendant vertex*, while a vertex of degree zero is an *isolated vertex*.
A graph H is shown in Fig. 4 where each vertex is labeled with its degree.

Fig. 4 A graph H where
each vertex is labeled with its
degree

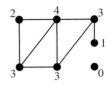

1.1   Families and Classes of Graphs

The structure of a graph can play a crucial role when studying graph theory. A *family*, or *class*, *of graphs* is a group of graphs that all have some similar type of structure. This structure is often characterized by one or more parameters: some value(s), which along with the rules for the graph, uniquely define it.

There are a large number of families of graphs. Let us look at some of the more common classes. Note that in graph theory, when discussing the number of vertices or edges, we always assume these values are nonnegative integers. Also, when talking about a "graph on n vertices," we mean that the graph contains n vertices (where $n > 0$).

A *cycle graph* on $n > 2$ vertices, denoted C_n, is a simple graph where every vertex has degree 2: for all v in the vertex set, $\deg(v) = 2$. It, as the name suggests, forms a circuit or cycle when drawn. Figure 5a shows an example of C_5.

The *path graph* on n vertices, denoted P_n, has two vertices of degree one and $n-2$ vertices of degree two. For $n > 2$, P_n can be constructed from C_n by removing one edge. A path graph on four vertices is given in Fig. 5b.

If a vertex is added to a cycle graph with n vertices where each vertex in the cycle is made adjacent to this new vertex, then we have a wheel graph. The n in W_n denotes the number of vertices on the cycle; W_n has $n + 1$ total vertices—the n vertices on the outer cycle, or rim, plus the center vertex, or hub. Formally, a *wheel graph* on $n > 2$ vertices, W_n, is a simple graph on $n + 1$ vertices with one vertex of degree n, called the hub, and n vertices of degree 3. An illustration of W_5 is given in Fig. 5c. We should note that some authors will choose to let W_n denote the wheel graph with n vertices (a rim of $n - 1$ vertices), so make sure you check the notation of the paper you are reading.

Another common class of graphs is the complete graphs, in which every vertex is adjacent to every other vertex in the graph. A *complete graph* on n vertices, K_n, has edge set $E(K_n) = \{uv \mid u, v \in V(K_n), u \neq v\}$. That is, the edge set of K_n consists of all distinct pairs of vertices. An example of K_8 is given in Fig. 6.

Exercise 2 Determine the degree of each vertex, as well as the total number of edges in terms of n, in the complete graph on n vertices, K_n.

To reiterate, n often represents the number of vertices in a graph. This is a convention we will use throughout this chapter when it is clear from the context that for some graph G, $|V(G)| = n$. However, there are some classes for which this is an exception. In the next family of graphs introduced below, the context is not so

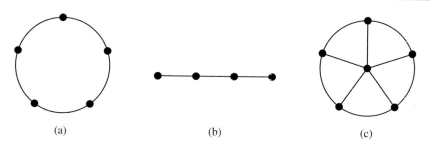

(a)   (b)   (c)

Fig. 5 Examples of classes of graphs. Note P_4 can be obtained from C_5 by removing a vertex and its incident edges, while W_5 can be obtained from C_5 by adding a central hub vertex that is connected to the original five vertices. (**a**) The cycle graph on five vertices, C_5. (**b**) The path graph on four vertices, P_4. (**c**) The wheel graph on six vertices, W_5

Fig. 6 The complete graph on eight vertices, K_8

straightforward and so we clearly state the *order*, or the number of vertices, of the graph.

A *bipartite graph* is a graph whose vertex set is composed of two disjoint sets X and Y such that every edge in the graph is incident to one vertex in X and one vertex in Y. In other words, no two vertices in X are adjacent, and no two vertices in Y are adjacent. An example of a bipartite graph is given in Fig. 7a. Combining the ideas of a bipartite graph with a complete graph, a *complete bipartite graph* on $n + m$ vertices is a graph whose vertex set consists of two disjoint set X and Y with $|X| = m$ and $|Y| = n$ such that for any $x \in X$ and for any $y \in Y$, xy is an edge in the graph. We denote this complete bipartite graph as $K_{m,n}$. The graph $K_{3,6}$ is illustrated in Fig. 7b. Also, the family of graphs $K_{1,n}$ are commonly referred to as star graphs and are also denoted as S_{n-1}. See Fig. 8b for a star on 10 vertices, $K_{1,9} = S_{10}$.

Exercise 3 How many edges are in the complete bipartite graph $K_{m,n}$? Write your answer in terms of m and n.

While a graph of the form $K_{1,n}$ is a star, it is also a tree. In graph theory, a *tree* is a connected graph that has no cycles. That is, a tree is a graph where there is one (and only one) path to travel from a vertex to any other vertex. A pendant vertex in a tree is called a *leaf*. A graph with n vertices is a tree if and only if it has $n - 1$ edges.

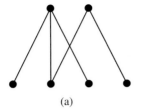
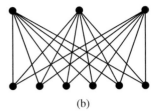

Fig. 7 Two bipartite graphs. The graphs are drawn so that the two partitions X and Y are separated: one at the top of the graph and one at the bottom. (**a**) A bipartite graph. (**b**) The complete bipartite graph $K_{3,6}$

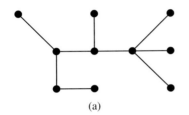

Fig. 8 Two different trees, each with 10 vertices. Like all trees, there are $n - 1 = 9$ edges, and in either case, if another edge were added to either graph, it would no longer be a tree. (**a**) A tree of order 10. (**b**) A star on 10 vertices

Exercise 4 Are the following graphs always, sometimes, or never trees? If the graph is sometimes a tree, define the values of n and/or m for which they are trees.

(a) C_n
(b) P_n
(c) W_n
(d) K_n
(e) $K_{m,n}$

One final class of graphs to introduce are those with Hamiltonian paths or Hamiltonian cycles. A *Hamiltonian path* in a graph is a path that visits each vertex exactly once. A *Hamiltonian cycle* (or *Hamiltonian circuit*) is a Hamiltonian path that is a cycle, meaning the starting and ending vertices are the same. A *Hamiltonian graph* is a graph which contains a Hamiltonian cycle. See Fig. 9 for examples. Clearly, all cycle graphs are Hamiltonian, given their cyclic structure.

Exercise 5 Which of the following graphs are Hamiltonian?

(a) W_n
(b) K_n
(c) S_n

Exercise 6 For what values of m and n is $K_{m,n}$ Hamiltonian?

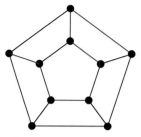
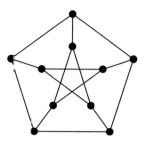

(a) A Hamiltonian graph. A Hamiltonian cycle is shown in red.

(b) The Petersen graph, an example of a non-Hamiltonian graph.

Fig. 9 In (**a**), we see an example of a Hamiltonian graph, while the graph in (**b**) is not Hamiltonian. However, the graph in (**b**) has a Hamiltonian path. Can you find one?

1.2   Graph Operations

As we saw in the descriptions of the class of path graphs and wheel graphs (or as we saw in Fig. 5), we can take a graph and apply some type of operation to it to obtain a different graph. In this section, we focus on taking two graphs and combining them in different ways to form a new graph. A word of warning: while the names of the operations are consistent throughout all graph theory literature, the notation is not! Always make sure to check the type of operation in a paper by reading the words and comparing it to the notation.

The disjoint union of two graphs G and H is created as the name suggests: form a graph with one copy of G and one copy of H with no additional changes. Formally, we have the definition below:

Definition 1 The *disjoint union* of graphs G and H, denoted $G \cup H$, is the graph with vertex set $V(G \cup H) = V(G) \cup V(H)$ and edge set $E(G \cup H) = E(G) \cup E(H)$.

Of course, we may take the disjoint union of more than two graphs. In this case, we often use G_i to represent our graphs, where $1 \leq i \leq k$, and we write $G_1 \cup G_2 \cup \ldots \cup G_k$ for the disjoint union of these k graphs. If this union consists of k copies of the same graph G, we will write kG instead of $\underbrace{G \cup G \cup \ldots \cup G}_{k \text{ times}}$.

The join of two graphs G and H takes the opposite approach to adding edges than the disjoint union does. This operation adds an edge between *every pair* of vertices whose endpoints are in different graphs (Fig. 10).

Definition 2 The *join* of graphs G and H, $G + H$, is the disjoint union of G and H along with all edges of the form $\{xy \mid x \in V(G), y \in V(H)\}$.

The join $P_4 + K_4$ is illustrated in Fig. 11.

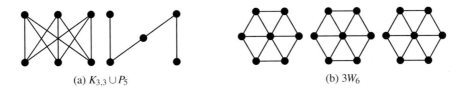

(a) $K_{3,3} \cup P_5$ (b) $3W_6$

Fig. 10 Graphs that are disjoint union of several well-known graphs. In (**a**), this graph is composed of a copy of $K_{3,3}$ and a copy of P_5. In (**b**), the graph is made up of three copies of W_6

Fig. 11 A representation of $P_4 + K_4$. Here, P_4 is highlighted in blue at the top, and K_4 is drawn in red at the bottom. The remaining edges come from the join operation

Exercise 7 Given the above definition of the join of graphs:

(a) Redefine W_n in terms of the join of two graphs that were introduced in Sect. 1.1.
(b) What other classes of graphs (C_n, P_n, K_n, $K_{m,n}$) can be defined using the join operation? Give the new definition or explain why the join operation cannot be used to define this family.

The last binary graph operation we will consider is the Cartesian product of two graphs. Just like graphing a point in the Cartesian plane, each vertex will be assigned a pair with the first coordinate from the first graph and the second coordinate from the second graph. Edges will only appear in this graph if one of the components in the pair is unchanged, but the other corresponded to an edge in either G or H.

Definition 3 The *Cartesian product* of G and H, written $G \times H$, is the graph with vertex set $V(G \times H) = V(G) \times V(H)$, and (x, y) is adjacent to (x', y') if and only if either $x = x'$ and $yy' \in E(H)$ or $y = y'$ and $xx' \in E(G)$.

An example of a Cartesian product involving two cycles is given in Fig. 12.
The m by n *grid* is the Cartesian product $P_m \times P_n$. If one of m, n equals two, say $n = 2$, we have the n-*ladder graph*, $P_n \times P_2$. This graph is often denoted L_n.

Exercise 8 Determine the order and number of edges for $P_m \times P_n$.

There are many other graph operations. However, the ones listed in this section provide the foundation and terminology needed to begin research on graph labelings.

Fig. 12 The Cartesian product of two cycles, C_3 and C_5. To better understand this construction, $G = C_3$ is shown in red to the left, and $H = C_5$ is shown in blue above $G \times H$

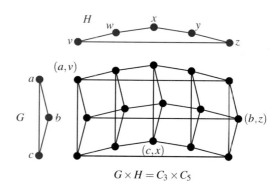

$G \times H = C_3 \times C_5$

1.3   Introduction to Graph Labeling

A *graph labeling* is just an assignment of some set of symbols (usually positive integers) to the vertices, edges, or both of a graph. More recently, labels on the faces of planar graphs have also been explored (for more about planar graphs and faces, see [24]). We have already seen a basic labeling in Fig. 4, which labeled each vertex with its degree. Typically, however, the idea of graph labeling stretches further than just assigning a defined value to the vertices and/or edges. The concept of a graph labeling also draws in some other requirement that must be satisfied by these assignments. For examples, perhaps adjacent vertices cannot have consecutive numbers as labels or perhaps edges must be labeled with the sum of the labels of their vertex endpoints. Satisfying these additional conditions is the key to creating a proper graph labeling. In this chapter, we focus on variations of *vertex graph labelings*, which, like outlined above, assign numerical values to only the vertices of a graph, subject to certain constraints (Fig. 13)

Labelings were first introduced in the late 1960s by Rosa [17] and have a broad range of applications, including communication networks, coding and cryptography, and optimization problems. We have composed an overview of one type of graph labeling that has many different questions: prime labelings. For all the different variations of prime labelings, we guide the reader through examples, provide exercises and challenge questions, and then present the reader with research questions, many of which are accessible to advanced high school students and early level undergraduate mathematics students. Though we have chosen to focus on prime labelings here, there are hundreds of other variations of labelings that you may explore. To learn more about the different types of graph labelings, consider investigating [11].

2   Coprime and Prime Labelings

The idea of a *prime labeling* of a graph originated from Etringer's work with trees around 1980 [9], though this concept was not formally introduced until Tout,

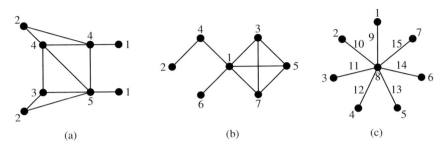

Fig. 13 Three different graphs with three different labeling rules. Note that (a) is asking to label each vertex with its degree (see Fig. 4), while (c) is labeling both vertices and edges. Typically, a rule is given as a definition, as we will see below. (**a**) A graph G labeled by the rule "each vertex is labeled with how many neighbors it has." (**b**) A graph H labeled with the rule "use 1 to 7 to label the vertices so that no adjacent vertices have consecutive numbers." (**c**) The graph S_8 labeled with the rule "label the vertices and edges with 1 to 15 so that the edges are labeled with the sum of the incident vertices"

Dabboucy, and Howalla published a paper [20] in 1982. In order to understand what a prime labeling is, we must first define what it means for two integers to be relatively prime.

Definition 4 Two integers are *relatively prime* (or *coprime*) if 1 is the greatest integer that divides them both.

In other words, two integers are *relatively prime* if and only if the greatest common divisor between them is 1. Some examples of relatively prime integers are 1 and 7; 28 and 29; and 24 and 49. However, 24 and 50 are not relatively prime because they share at least one factor greater than 1 (in this case, just the integer 2). Understanding when some integers are relatively prime is important in understanding prime labelings.

Exercise 9 Provide an argument that $\gcd(1, n) = 1$, where n is an arbitrary positive integer.

Exercise 10 Above, we gave the example of 28 and 29 being relatively prime integers. Determine, with explanation, whether two consecutive positive integers are always relatively prime.

Though relatively prime is the most common term for two integers whose greatest common divisor is 1, some may refer to these integers as *coprime*.

Definition 5 A *coprime labeling* of a graph G of order n is a labeling in which adjacent vertices are given distinct relatively prime labels from the set $\{1, 2, \ldots, k\}$, for some $k \geq n$.

Fig. 14 A graph G with a coprime labeling from $\{1, 2, \ldots, 9\}$

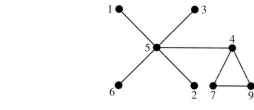

Fig. 15 A graph with a prime labeling [6]

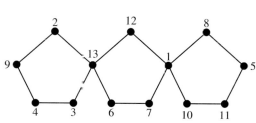

An example of a coprime labeling is given in Fig. 14. Observe that no two vertices have the same label, and the greatest common divisor of the labels of the endpoints of edges is 1.

Exercise 11 Consider the graph G given in Fig. 14.

(a) Find a different coprime labeling of this graph using the same set $\{1, 2, \ldots, 9\}$ or explain why no such labeling is possible.

(b) Find a coprime labeling using $\{1, 2, \ldots, 8\}$ or explain why no such labeling is possible.

(c) Find a coprime labeling using $\{1, 2, \ldots, 7\}$ or explain why no such labeling is possible.

Prime labelings are a special case of coprime labelings.

Definition 6 A graph has a *prime labeling* if the vertices of G can be labeled distinctly with the integers $\{1, 2, \ldots, n\}$, where n is the number of vertices, such that any two adjacent vertices are relatively prime. A graph is said to be *prime* if it has a prime labeling.

Notice that you are asked to find a prime labeling in part (b) of Exercise 11 above.

An example of a graph and its prime labeling appears in Fig. 15. The graph has 13 vertices, and each vertex is labeled with the integers between 1 and 13, inclusive. Per the definition, the greatest common divisor of any two vertices that share an edge is 1.

Exercise 12 Consider the graph G given in Fig. 16. Answer the questions below:

Fig. 16 The graph G for
Exercise 12

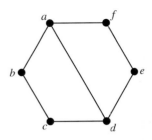

(a) Let the labels for vertices a, b, and c be 1, 2, and 3, respectively. Can you assign
the remaining labels so that G is prime? If so, give all possible such labelings,
otherwise explain why no such prime labeling is possible.
(b) Let the labels for vertices b, c, and d be 1, 2, and 3, respectively. Can you assign
the remaining labels so that G is prime? If so, give all possible such labelings,
otherwise explain why no such prime labeling is possible.
(c) Ignoring labelings that are the same under rotations and reflections, are there
any other ways to label G so that it is a prime graph? If so, list these labelings,
otherwise explain why no additional labelings exist.

After thinking about the definitions of a graph having a coprime and a prime
labeling, it may seem clear that if a graph has a prime labeling, then it has a coprime
labeling. We will formally state and prove this idea using a direct proof in the
following exercise. When beginning to construct a proof, it is important that you
always do a few things:

1. Begin by stating what you are assuming about the graph. This could be the
number of vertices, the number of edges, the type of graph that it is, or some
other defining information.
2. Refer or restate definitions as you use them, and then make sure you use them
properly.
3. Write in complete sentences and make sure each sentence follows logically from
the one before.
4. Conclude by stating the final result. In other words, say you did exactly what the
problem stated.

Exercise 13 For the proposition and proof below, fill in the blanks to create a solid,
logical argument.

Proposition 1 *For a graph G on n vertices, if G has a _____ labeling, then G has
a _____ labeling.*

Proof Let G be a graph of order _____. Assume G is a prime graph. That is,
assume G _____(use definition here)_____.

For G to have a coprime labeling of G, we must label _____with the numbers _____, for some k_____n. But we know G is prime, so we can label _____with _____, and, since $n \leq k$, we must have G can be labeled with the set _____. Hence, G has a _____ labeling, as we set out to show. □

Proofs are the backbone of pure mathematics: they provide solid, logical arguments for why a certain statement is true. For a quick guide of writing proofs, we encourage you to refer to [13]. For a deeper understanding of logic and proofs, consider reading *Mathematical Reasoning: Writing and Proof* [19], a free open source textbook, or *Mathematical Proofs: A Transition to Advanced Mathematics* [7]. Remember, to prove that a graph is prime, all you need to do is to find one prime labeling. You can do this by providing a generic labeling that would work for any graph in that family. We will guide you how to do this for paths and cycles in the following exercises.

Exercise 14 We will prove that all paths are prime by answering the following questions:

(a) First, convince yourself that P_n is prime and then find a prime labeling. Do this by the following steps:
  (i) Draw and label P_n for different values of n such as P_3, P_4, P_5, ... so that P_n has a prime labeling. Focus on creating a pattern of label assignments.
  (ii) Then, draw and label P_n, for an arbitrary n, so that P_n is prime. (Hint: for each example in (a), write "n" instead of a number. How can you rewrite other labels with the letter n?)
(b) Write your mathematical proof by doing the following:
  (i) Write the statement we will prove. Be exact, making sure you include n in the statement.
  (ii) Begin the proof by writing what you know about the graph and the set of allowable labels.
  (iii) Argue your labeling.
  (iv) Conclude by stating what has been proven.

Exercise 15 Adapt Exercise 14 to prove that all cycles are prime.

There are several different prime labelings for the paths and cycles. For the path graph, P_n, you may have simply labeled the vertices along the path consecutively. Because any two adjacent vertices will be labeled consecutively (and consecutive integers are relatively prime), this would result in a prime labeling of the path. The same labeling will also work for cycle graphs. The only two vertices that will not be consecutively labeled are the first vertex (labeled 1) and the last vertex (labeled n). Because 1 and n are relatively prime, this labeling is prime. Again, there are several other labelings that would also work.

Exercise 16 Can you find another prime labeling for the graph in Fig. 15?

Fig. 17 The graph
$G = C_5 \cup C_9$

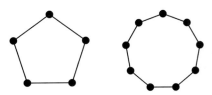

Recall, from Definition 1, that the disjoint union of graphs is an operation that combines two or more graphs to form a larger graph by making the vertex set of the result be the disjoint union of the vertex sets of the given graphs and by making the edge set of the result be the disjoint union of the edge sets of the given graphs. Figure 17 gives a disjoint union of the cycles C_5 and C_9.

Example 1 The graph $C_5 \cup C_9$ is not prime.

This is the first time we have encountered a graph that *is not* prime. Oftentimes, one will argue that a graph is not prime by contradiction—meaning that we explore what would happen if it *were prime* and then finding a contradiction of something we know to be true. (To learn more about proofs by contradiction, see [19] or [7]).

Exercise 17 Write a formal proof that the graph given in Fig. 17 is not prime by filling in the blanks in Proposition 2 below.

Proposition 2 *The graph* _____ *is not prime.*

_____. (Begin by stating what you are assuming) . Let us proceed by contradiction. That is, assume G =_____ is prime. Since G has a prime labeling, G (insert definition here) .

Now, in the set of labels $\{1, 2, \ldots, 14\}$, there are _____ even integers. The cycle C_5 can be labeled with at most _____ even integers in order for this cycle to have a prime labeling. Similarly, _____ can have at most _____ even integers labels in any prime labeling of this graph.

Since $2 + 3 = 5$ and _____ < _____, not all of the even integers from $\{1, 2, \ldots, 14\}$ are used as labels. Thus, we have arrived at a contradiction to G being prime, and our original assumption is _____.

Therefore, _____.   □

Essentially, if $G = C_5 \cup C_9$ were prime, then the vertex labels must be integers 1 through $5 + 9 = 14$. Therefore, the vertex labeling must use $\frac{14}{2} = 7$ even integers as labels. This means that at least half of the vertices from one of the cycles, either C_5 or C_9, must be labeled with even integers. Since each cycle has an odd length, two adjacent vertices would be labeled with even integers, contradicting that these two vertex labels are relatively prime.

Exercise 18 Prove $2C_{11}$ is not prime.

In Exercise 18, mimic the ideas and structure from Exercise 17.

Challenge Problem 1 Let $G = C_{n_1} \cup C_{n_2} \cup \cdots \cup C_{n_k}$. Prove that G is not prime when at least two n_i are odd. (This is shown in Theorem 2.1 of [10].)

Exercise 19 Explain why the graph K_n is not prime when $n \geq 4$. As a challenge, write this as a proposition and prove it.

Other graphs have also been shown to be prime. Berliner et al. [5] studied the prime labelings of ladder graphs and complete bipartite graphs. We have listed some of their results as exercises below.

Exercise 20 Find a prime labeling for $K_{3,n}$ where $n \neq 7$. (This is shown in Proposition 20 of [5].)

Exercise 21 Prove that if $n + 1$ is prime, then the ladder graph $P_n \times P_2$ has a prime labeling. (This is proven in Theorem 1 in [5].)

Challenge Problem 2 Prove that if n is even and the wheel graph W_n is prime, and if n is odd, then W_n is not prime. (Combined these, prove that W_n is prime if and only if n is even, which is shown in [14].)

The primeness of several other graphs formed from cycles, such as graphs known as prisms and the generalized Petersen graph, was also explored in [18] (Fig. 18).

2.1   Minimal Coprime Labeling

As you showed earlier in this section, many graphs do not have prime labelings; however, because there are infinitely many prime numbers, **every** finite graph must have a coprime labeling because each vertex can simply be labeled with a different prime number. Therefore, we define $pr(G)$ to be the minimum value of k for which graph G has a coprime labeling, and the corresponding labeling of G is called a *minimal coprime labeling* of G. If G is prime, then $pr(G) = n$.

When showing $pr(G) = k$, you need to do two things: (1) find a coprime labeling of G using integers from $\{1, 2, \ldots, k\}$ and (2) argue that there is not a coprime

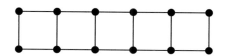

Fig. 18 The ladder graph, $P_6 \times P_2$. Note that the solution to Exercise 21 will show that this graph is prime

labeling from $\{1, 2, \ldots, k-1\}$. In other words, you must show that using the number k to label a vertex is necessary.

Exercise 22 We know that when n is even $\text{pr}(W_n) = n + 1$ because in this case, W_n is a prime graph. Show that, when n is odd, $\text{pr}(W_n) = n + 2$. (This is shown in Proposition 2 of [2].)

Exercise 16 shows that prime labelings of graphs are not necessarily unique. However, research has often centered around whether a graph is prime, and not in the number of unique labelings, though enumerating these possibilities is certainly a possible avenue for further research.

Research Project 1 How many unique prime labelings exist for a specific class of prime graphs? To get started on this question, you may consider exploring even and odd length cycles or other classes of graphs, such as wheels and stars. If you cannot find an exact value, can you find a lower bound?

3   Consecutive Cyclic Prime Labelings

The idea of a *consecutive cyclic prime labeling* was first introduced by Berliner et al. in [5]. However, it was specifically defined for ladder graphs. The following is Definition 2 in [5].

Definition 7 A *consecutive cyclic prime labeling (CCPL) of a ladder* $P_n \times P_2$ is a prime labeling in which the labels on the vertices wrap around the ladder in a consecutive way. In particular, if the label 1 is placed on vertex v_i, then 2 will be placed on v_{i+1}, $n - i + 1$ on v_n, $n - i + 2$ on u_n, $2n - i + 1$ on u_1, $2n - i + 2$ on v_1, and $2n$ on vertex v_{i-1}. A similar definition holds if 1 is placed on u_i.

Exercise 23 Construct a CCPL of $P_n \times P_2$ for $2 \le n \le 6$.

Now, to show a graph has a consecutive cyclic prime labeling (CCPL), we need to show that (1) the vertices are labeled consecutively and (2) the labeling is prime. Now, because ladder graphs have a Hamiltonian cycle that starts at v_i and ends at v_{i+1}, then to show a ladder has this CCPL, we need to answer: does labeling the vertices consecutively along this ladder yield a prime labeling? A characterization for the ladders with consecutive cyclic prime labeling is provided in [5]. We divide this characterization into two exercises for the reader below (Fig. 19).

Exercise 24 If $P_n \times P_2$ has a consecutive cyclic prime labeling with the value 1 assigned to vertex v_1, then $2n + 1$ is prime. (Hint: use contradiction.)

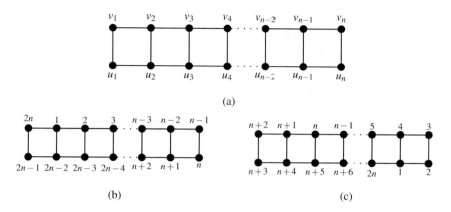

Fig. 19 The ladder $P_n \times P_2$ with the two consecutive cyclic prime labelings given in Definition 7. (a) The ladder $P_n \times P_2$ with vertices identified. (b) $P_n \times P_2$ labeled as in Definition 7 with $v_i = v_2$ labeled as 1. (c) $P_n \times P_2$ labeled as in Definition 7 with u_{n-1} labeled as 1

Now, you will consider the converse of Exercise 24. Note that when proving the converse of a statement what you were assuming is what you now need to prove and what you were proving is now your assumption at the start of the proof.

Exercise 25 If $2n+1$ is prime, then $P_n \times P_2$ has a consecutive cyclic prime labeling when the value 1 is assigned to vertex v_1.

Putting the results of Exercises 24 and 25 together, what can we say about ladder graphs with consecutive cyclic prime labelings?

Theorem 1 $P_n \times P_2$ *has a consecutive cyclic prime labeling with the value 1 assigned to vertex v_1 if and only if $2n + 1$ is prime.*

In [10], Donovan et al. adapted the definition of a cyclic prime labeling of ladders and applied it to the disjoint union of cycles. When labeling the cycles, they label each graph both *consecutively* (the labels appear in order) and *cyclically* (the labels wrap around the cycle).

Definition 8 For a cycle on n vertices, $v_1, v_2, \ldots, v_n, v_1$, the consecutive cyclic labeling φ is defined as $\varphi(v_{i+1}) = \varphi(v_i) + 1$, where $v_{n+1} := v_1$. Any consecutive cyclic labeling resulting in a prime graph is then a consecutive cyclic prime labeling.

Definition 8 is illustrated in Fig. 20.
The work [10] focuses on the existence of consecutive cyclic prime labelings of the disjoint union of cycle graphs. Recall from Definition 1 that the disjoint union of two graphs is one copy of G and one copy of H with no additional changes. The disjoint union of cycles therefore consists of several independent cycles. In this

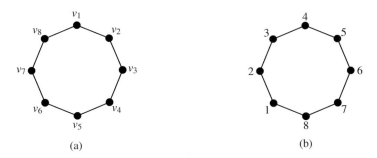

Fig. 20 A cycle on eight vertices as well as a consecutive cyclic labeling of this graph. Is the labeling given in (b) prime? (**a**) C_8 with vertices identified. (**b**) C_8 with a consecutive cyclic labeling. Note here that $\varphi(v_6) = 1$

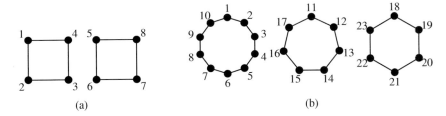

Fig. 21 Two examples of disjoint unions of graphs. Could either of these labelings be extended by adding a copy of C_4? A copy of C_5? (**a**) $2C_4$ with a CCPL. (**b**) $C_{10} \cup C_7 \cup C_6$ labeled with a CCPL

case, the cycles are labeled consecutively and cyclically in the given order. From Challenge Problem 2, the disjoint union of odd cycles is not prime. Thus, we know it will not be consecutive cyclic prime. However, are there disjoint unions of cycles that are consecutive cyclic prime? See Fig. 21 for two examples.

Perhaps we first consider the union of identical cycles, mC_n. The question at hand becomes: how large can m be so that a CCPL of mC_n remains prime? Now, if n is odd, we know mC_n is prime only when $m = 1$. Because a consecutive cyclic labeling of one odd cyclic is also prime, then, mC_n is consecutive cyclic prime *only when* $m = 1$ for odd n. However, how many copies of even cycles result in a consecutive cyclic prime labeling?

Exercise 26 Show that $3C_6$ has a consecutive cyclic prime labeling.

Exercise 27 Is $4C_6$ consecutive cyclic prime? is $5C_6$? In either case, explain.

What we are discovering is the upper limit on the possible number of cycles we may have in a disjoint union for which the graph is prime when labeled with the consecutive cyclic labeling φ from Definition 8. In Exercises 26 and 27, we observed that each pair of adjacent vertices in each cycle is consecutive integers with the exception of the first and last vertices labeled in the cycle. Recall that consecutive integers are relatively prime; therefore, we only need to check the first and last labeled vertices in each cycle for common factors.

Still focusing on mC_n, the kth copy of C_n has its first vertex labeled with $(k-1)n+1$ and its last is labeled with kn, as we move around the cycle. We need the result of the following exercise to continue this discussion.

Exercise 28 Show that a common factor of two integers must also divide the difference of those integers. That is, prove the following: for integers m and n with $gcd(m, n) = d$, $m - n$ is a multiple of d.

Thus, by Exercise 28, a common factor d of these labels must divide $kn - ((k-1)n+1) = n-1$. This results in the following lemma, which appears in [10].

Lemma 1 *Any common factor of two vertex labels in a cycle of length n labeled via a consecutive cyclic labeling must be a factor of $n - 1$.*

Lemma 1 is very useful when trying to determine whether the disjoint union of cycles of the same length is consecutive cyclic prime. Use Lemma 1 to prove the three statements in Exercise 29 that compose Theorem 2.2 in [10]. Recall that m is the number of copies of C_n in the disjoint union of the cycles.

Exercise 29 Let $G = mC_n$ for some $n \geq 4$, and let p be the smallest prime factor of $n - 1$.

(a) There exists a consecutive cyclic prime labeling of mC_n for $1 \leq m \leq p - 1$.
(b) The maximum possible m for which G is prime with this labeling is $m = p - 1$.
(c) For the special case where $n - 1$ is itself prime, $m \leq n - 2$, and if n is odd, $m = 1$.

Because consecutive cyclic prime labelings were first only introduced for ladder graphs, we wonder whether this definition can be generalized for other graphs with either Hamiltonian cycles or paths, similar to what was done in [10]. We adapt the definition below.

Definition 9 Let G be a graph on n vertices with Hamiltonian cycle $v_1, v_2, \ldots, v_n, v_1$. Then, a consecutive cyclic labeling of G would label the ith vertex of the cycle with the integer i. In other words, $\varphi(v_i) = i$.

Research Project 2 Which Hamiltonian graphs are consecutively cyclic prime?

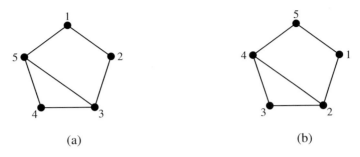

Fig. 22 A graph G consisting of a 5-cycle with an additional edge. Note the role the labeling plays here: in (**a**), G is prime, but in (**b**) it is not

When answering this question, it is helpful to remember that any vertex in a Hamiltonian cycle can be considered the first. For example, Fig. 22 shows two consecutive cyclic prime labelings of the same graph. However, only the second labeling is prime.

You may also consider the disjoint union of Hamiltonian graphs, as [10] did with cycles. Since [5] showed which ladder graphs have consecutive cyclic prime labelings, you may consider starting with disjoint unions of ladders, all of which have the same size. Then, you can expand upon your results and consider other graphs.

Research Project 3 If G is a Hamiltonian graph with a consecutive cyclic prime labeling, what is the maximum m so that mG has a CCPL?

You can also adapt Definition 9 to be for graphs with Hamiltonian paths, not Hamiltonian cycles, which may provide you with more flexibility.

Definition 10 Let G be a graph on n vertices with Hamiltonian path $v_1, v_2, \ldots, v_n$. Then, a consecutive cyclic labeling of G would label the ith vertex of the path with the integer i. In other words, $\varphi(v_i) = i$.

Research Project 4 Which graphs with Hamiltonian paths are consecutively cyclic prime?

Finally, perhaps you will consider adapting the definition of a coprime labeling to cyclic coprime labelings. As we saw in Sect. 2, all graphs have a coprime labeling. A *cyclic coprime labeling* would be a labeling that labels the vertices of a Hamiltonian cycle with increasing integers, instead of (increasing) consecutive integers.

Definition 11 Let G be a graph on n vertices with Hamiltonian cycle $v_1, v_2, \ldots, v_n, v_1$. Then a cyclic labeling of G would label the vertices of the cycle with increasing integers. In other words, $\varphi(v_{i+1}) > \varphi(v_i)$ where $v_{n+1} := v_1$.

Research Project 5 What is the minimum value of k for which a Hamiltonian graph G has a cyclic coprime labeling?

You may also consider introducing the idea of a *strongly consecutively cyclic prime graph*. To our knowledge, this has not yet been introduced, but it would be in line with the idea of a strongly prime graph. A graph is a *strongly prime graph* if for any vertex v of G there exists a prime labeling for which v is labeled with 1. The idea of a strongly prime graph was introduced by Vaidya and Prajapati in [22].

Research Project 6 What graphs, if any, have a strongly consecutively cyclic prime labeling? In other words, what graphs have a consecutively cyclic prime labeling for which *any* vertex of G can be labeled with 1?

We know that the graph in Fig. 22 would not be strongly consecutively cyclic prime. However, what graphs, other than cycles, would be?

These are not the only directions to take for future research. Remember that mathematicians often adapt definitions or introduce new definitions when doing research.

4   Neighborhood-Prime Labelings

What we have seen thus far is that interesting and significant mathematical results have been made on various alterations of prime labelings. In this chapter, we discuss our final variation—neighborhood-prime labelings. Neighborhood-prime labelings were first introduced by Patel and Shrimali [16] The neighborhood of a vertex is the set of all vertices adjacent to that vertex. This set of neighbors does not contain the vertex itself. If the vertices of a graph can be labeled with $\{1, 2, \ldots, n\}$ so that for any given vertex with degree greater than 1 the greatest common divisor of the neighborhood labels is 1, the graph is a *neighborhood-prime graph*. An example of such a graph is given in Fig. 23.

Exercise 30 Repeat Exercise 12, replacing "prime" with "neighborhood-prime."

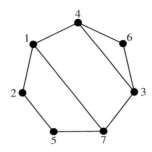

Fig. 23 An example of a graph with a neighborhood-prime labeling. Note that, for each vertex with degree of 2 or more, we must check the labeling by finding the greatest common factor of all the neighborhood labels; for the vertex labeled 1 with neighbors labels 2, 4, and 7, their greatest common divisor is 1

Fig. 24 Neighborhood-prime graphs G and G', where G is constructed by attaching the vertex labeled 6 to the vertex labeled 3. Note that, with the exception of the pendant vertex, this new vertex could be attached to any vertex and G' would still be neighborhood-prime. (**a**) A neighborhood-prime graph, G. (**b**) A neighborhood-prime graph, G' constructed by adding a pendant vertex to a vertex of degree 2

Exercise 31

(a) Let G be a graph with a neighborhood-prime labeling. Argue that the graph, G', resulting from attaching a pendent vertex to exactly one vertex of degree at least 2 in G is also neighborhood-prime. An example of such a graph appears in Fig. 24.
(b) Explain why the new vertex cannot necessarily be attached to a pendant vertex. How can the statement in part (a) be adjusted to allow attachment to a pendant vertex and have G' be neighborhood-prime?

   Exercise 31 shows one way we can build upon neighborhood-prime graphs to create new neighborhood-prime graphs with more vertices. However, does the same idea work for edges?

Exercise 32 Let G be a graph with a neighborhood-prime labeling. Consider the graph $G+e$, the graph formed by adding an edge between two non-adjacent vertices, u and v, in $V(G)$. What must be true about u and v to guarantee that $G + e$ is also neighborhood-prime?

As we established in Exercises 31 and 32, it is important that the vertices to which we are adding have degree greater than or equal to 2 to guarantee that the new, larger graph is also neighborhood-prime.

You may have been curious whether all prime labelings of graphs are also neighborhood-prime labelings.

Exercise 33 Determine whether all prime labelings of a graph are also neighborhood-prime labelings. Explain your reasoning.

There are several examples in this chapter thus far that show that the answer is no. In Sect. 2, we discussed that labeling all vertices on a path consecutively yields a prime labeling. However, this labeling is not a neighborhood-prime labeling for paths with more than four vertices. This is because the vertex labeled with 3 will be adjacent to vertices only labeled with 2 and 4, which are not relatively prime integers. However, every path does have a neighborhood-prime labeling.

Exercise 34 Prove that path P_n has a neighborhood-prime labeling for every n.

Similar to paths, the consecutive labeling discussed in Sect. 2 does not work for cycles. Can we find a variation for odd cycles?

Exercise 35 Prove that cycles of odd length, C_n, where n is odd, have a neighborhood-prime labeling.

Before we continue, it will be helpful for us to establish a basic understanding of modular arithmetic. The remainder when two integers are divided will be key in understanding the results given below. Now, we know that 13/3 has a remainder of 1 (since $13 = 4*3+1$), so, using the idea of modular arithmetic, we can write $13 \equiv 1$ (mod 3). Similarly, 11/3 can be written as $11 \equiv 2$ (mod 3) since $11 = 3*3+2$. We will briefly explore the idea of mods in Exercise 36 below. More detail can be found in almost any discrete mathematics or number theory text. However, we recommend first exploring Chapter 5 in [15], an open-source text.

Exercise 36

(a) What are the possible remainders when an integer is divided by 4? by 5? by k?
(b) List five numbers n that satisfy $n \equiv i$ (mod 4) for $i = 0, 1, 2, 3$.
(c) What does it mean for a number k to be equivalent to 0 (mod n)?
(d) What is the smallest positive integer n such that $n \equiv 0$ (mod k), where $k > 1$?
(e) What is the smallest positive integer n such that $n \equiv 1$ (mod k), where $k > 1$?

Now, build on your results from Exercise 35 and use the ideas introduced above to show the following:

Exercise 37 Show that the cycle C_n is neighborhood-prime if $n \not\equiv 2$ (mod 4).

This leaves the question, "but what about C_n if $n = 2$ (mod 4)?" for example, cycles of length 6, 10, 14, and so on. Well, it turns out that a neighborhood-prime labeling for those graphs *does not* exist.

Challenge Problem 3 Prove that the cycle C_n is not neighborhood-prime when $n \not\equiv 2$ (mod 4). (This is proven in Theorem 2.7 of [16].)

Putting the results of the exercises above together results in the following theorem.

Theorem 2 *The cycle C_n has a neighborhood-prime labeling if and only if $n \not\equiv 2$ (mod 4).*

You can apply the result from Exercise 32 to Theorem 2 and obtain something even more powerful, which was done in [3].

Theorem 3 *Let G be a graph of order n such that $n \not\equiv 2$ (mod 4). If G is Hamiltonian, then G has a neighborhood-prime labeling.*

Exercise 38 Prove Theorem 3.

4.1 Building on Cycles

The results from Exercises 31 and 32, as well as Theorem 3, show that one can build larger graphs from graphs that are known to be neighborhood-prime.

Much of the existing research in the field of neighborhood-prime graphs deals with graphs containing cycles or that are constructed from graph operations with cycles. Many results can be found in [3, 8], and [1]. In each of these papers, the authors find new neighborhood-prime graphs by building on existing neighborhood-prime graphs. We will choose to focus on the results on two classes of graphs, book graphs, and stacked prisms, in the following exercises.

Let us first consider the family of book graphs. An *m-book* graph is a graph in which m 4-cycles all share a common edge. This graph is the Cartesian product of the star S_{m+1} and P_2. There are several variations of book graphs, but here we will focus on the k-polygonal book graphs, a generalization of the m-book graph. The *k-polygonal book*, denoted B_n^k, is formed by n copies of a k-polygon C_k, all sharing a single edge called the *spine*. We denote the two endpoints of the spine as s_1 and s_2. We refer to each k-polygon, C_k, as a page of the k-polygonal book graph. Figure 25 shows an example of B_3^5.

Exercise 39 Find a neighborhood-prime labeling of B_3^k for $k = 4, 5, 6$. See Fig. 25 for an illustration of B_3^5.

Fig. 25 The 5-polygonal book, B_3^5

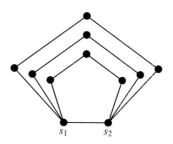

Now, build on your results from the exercise above to do the following exercise.

Exercise 40 Prove that all 5-polygonal books are neighborhood-prime. (This is done in Theorem 9 of [8]).

Now that we have generalized neighborhood-prime labelings of book graphs for a specific size polygon (in this case, a polygon of size 5), let us consider generalizing this type of labeling to a specific number of pages

Exercise 41 Prove that all k-polygonal books with 3-pages, B_3^k, are neighborhood-prime.

Challenge Problem 4 Prove that all k-polygonal books B_n^k where $k \geq 3$ and $k \equiv 0$ (mod 4) are neighborhood-prime. (This is proven in Theorem 4.1 of [1].)

Challenge Problem 5 Prove that if $k \equiv 2$ (mod 4), then B_n^k is not neighborhood-prime. (This is proven in Theorem 4.2 of [1].)

A book graph is a special class of stacked prisms. *Stacked prisms* are the Cartesian product of a cycle with a path of length n. More formally, the stacked prism graph $Y_{m,n}$ is the graph Cartesian product $C_n \times P_n$. These graphs can be drawn as multiple concentric identical cycles where each vertex in a cycle is adjacent to the equivalent vertices in the cycles immediately within it and immediately beyond it An example from [1] is shown in Fig. 26.

Exercise 42 Find a neighborhood-prime labeling of $Y_{3,3}$.

Similar to how we did with book graphs, let us build on the above exercise and obtain more general results.

Exercise 43 Prove that all stacked prisms of the form $Y_{3,n}$ are neighborhood-prime.

Now, let us see if we can generalize our results in the following challenge problem.

Fig. 26 The stacked prism $Y_{6,3}$, obtained by the Cartesian product $C_6 \times P_3$

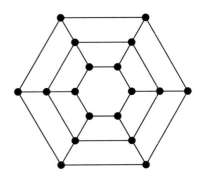

Challenge Problem 6 Prove that all stacked prisms $Y_{m,n}$ are neighborhood-prime. This is proven in Theorem 3.1 of [1].

If you are interested in research projects for neighborhood-prime labelings, we encourage you to consider building other neighborhood-prime graphs through graph operations (as we did above with some examples of Cartesian products). However, choose your Cartesian products carefully so that your results do not directly follow from Theorem 3. For example, the k-polygonal books are not Hamiltonian because of the spine of length two. However, the stacked prism graphs *are* Hamiltonian. However, Theorem 3 only tells us that Hamiltonian graphs of order $n \not\equiv 2 \pmod 4$ have a neighborhood-prime labeling, and it does not tell us about the case where $n \equiv 2 \pmod 4$. Thus, the result on all stacked prism graphs is original.

Research Project 7 What other neighborhood-prime graphs can be built through graph operations. For example, when is the join of graphs neighborhood-prime? When is the disjoint union of neighborhood-prime graphs also neighborhood-prime?

You may also consider building upon the result of polygonal book graphs, perhaps considering book graphs with a spine that contains more than one edge.

Research Project 8 Is there a general pattern for graphs to be neighborhood-prime that contain many identical cycles with a shared path of any length?

Finally, because many of the graphs explored here are Cartesian products of smaller graphs that are known to be neighborhood-prime, is the following true?

Research Project 9 Determine whether the following conjecture from [1] is true: if G and H both are neighborhood-prime graphs, then $G \times H$ is also neighborhood-prime.

4.2   Building on Trees

The previous section focused on building neighborhood-prime graphs using results on cycles. However, what if a graph does not contain a cycle? Recall that such graphs (when connected) are called trees. Only a few classes of trees have been investigated. The work [8] shows that all caterpillars, spiders banana trees, firecrackers, and bivalent-free trees are neighborhood prime. Let us explore a few of these below.

A *caterpillar graph* is a tree in which a single path (the spine) is incident to (or contains) every edge. That is, a caterpillar graph is a tree where the removal of every vertex of degree one leaves a path graph. An example of a caterpillar graph is given in Fig. 27.

Exercise 44 Use Exercise 31 to argue that all caterpillars are neighborhood-prime.

A *firecracker graph*, $F_{n,k}$, is a graph obtained from a chain of n k-stars by linking one leaf from each star along a path of length n. Two examples are shown below. Note that $F_{1,k}$ is just the star S_k, and $F_{n,1}$ is just a path. Also, $F_{n,2}$ is a caterpillar (Fig. 28).

Challenge Problem 7 Prove that $F_{n,k}$ is neighborhood-prime for all positive n and k. This is proven in Theorem 15 of [8]. Note: you only need to consider when $k \geq 3$ since the cases where $k = 1$ or $k = 2$ are included in other classes known to be neighborhood-prime.

A *bivalent-free tree* is a tree in which all vertices are leaves or have degree greater than or equal to three. In other words, a bivalent-free tree has no vertices of degree two (Fig. 29).

Challenge Problem 8 Prove that all bivalent-free trees are neighborhood-prime.

Cloys and Fox [8] showed that bivalent-free trees are neighborhood-prime in Theorem 17 of their paper. An alternative labeling, using a breadth-first search, is given in [1]. We bring up these two proofs because, as with prime labelings,

Fig. 27 A caterpillar whose spine is of length six

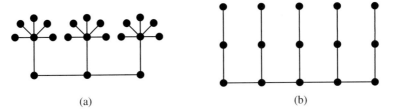

Fig. 28 Two examples of firecracker graphs, $F_{3,7}$ and $F_{5,3}$. Recall that the first index in the notation gives the length of the path, while the second gives the size of each star linked to this path. (**a**) The firecracker graph $F_{3,7}$. (**b**) The firecracker graph $F_{5,3}$

Fig. 29 A bivalent-free tree. Notice that each vertex has degree 1 or degree greater than or equal to 3

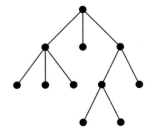

neighborhood-prime labelings are not necessarily unique. Oftentimes, one labeling may be easier than another to build larger graphs that are neighborhood-prime graphs, as was done in [1] with bivalent-free trees. The benefit of the breadth-first labeling algorithm is that *any* non-leaf vertex can be the root and thus labeled 1. This provides more flexibility than other labelings.

The work done on trees has led Cloys and Fox to conjecture that all trees are neighborhood-prime. This seems like quite a lot to dive into! Therefore, what other trees can we show are neighborhood-prime? Because we know bivalent-free trees are neighborhood-prime, we need to focus on classes of trees that contain vertices of degree two.

> **Research Project 10** What classes of trees that contain vertices of degree two are neighborhood-prime? Perhaps you may first explore trees with exactly one vertex of degree 2 and then build upon your work.

The work in [8] on caterpillars, which you also showed in Exercise 44, may lead one to first investigate *lobster graphs* or trees for which every vertex is within distance two of a central path. In other words, a lobster graph is a tree with the property that the removal of leaf vertices leaves a caterpillar. (To learn more about lobster graphs or to explore other classes of trees, we recommend exploring those listed in [23].)

> **Research Project 11** Are all lobster graphs neighborhood-prime?

We should note that we anticipate answering the question in Research Project 11 to be significantly more challenging than first exploring trees with a small number of vertices of degree 2.

4.2.1 Further Projects on Neighborhood-Prime Labelings

Below we present the reader with more potential research projects in the area of neighborhood-prime labelings. Remember that mathematicians often introduce their own definitions or adapt definitions when they conduct research.

> **Research Project 12** As we saw in Sect. 2, all graphs have coprime labelings. Therefore, you may investigate the minimum value for k for which G has a neighborhood-coprime labeling.

Akin to the idea of a strongly consecutively cyclic prime labeling defined in Sect. 3, perhaps you may consider introducing the idea of a strongly neighborhood-prime graph. While we do not think this idea has been previously introduced, it would be another extension of a strongly prime graph. A graph is a *strongly prime graph* if for any vertex v of G there exists a prime labeling for which v is labeled with 1. The idea of a strongly prime graph was introduced by Vaidya and Prajapati in [22].

> **Research Project 13** What graphs have a strongly neighborhood-prime labeling? Begin by writing an argument showing that paths and cycles are strongly neighborhood-prime, and then see what other results you find. Feel free to build off results that are already in the literature.

5 Conclusion

There are a variety of research projects presented in this chapter that are accessible for new mathematicians. Though several of these build off work that has already been done, many of the research problems adapt existing mathematical definitions to create new areas of research. If you wish to investigate research questions that have not been explored in the existing literature, we recommend you dive into Research Problems 2, 3, 4, 5, 6, 12, or 13. Moreover, for those who feel they need to strengthen their understanding of some concepts, we have referenced numerous easily accessible open-source textbooks. However, feel free to find texts that are better suited to your needs as you delve into the land of prime graphs. Happy label hunting!

References

1. Ablondi, T., Hayes, A. and Wiglesworth, L. W. Results on Neighborhood-Prime Labelings of Graphs. Preprint (2019).
2. Asplund, J., Fox, B.N.: Minimum Coprime Labelings for Operations on Graphs. Integers. **19** (2019).
3. Asplund, J., Fox, N. B., Hamm, A. New Perspectives on Neighborhood-Prime Labelings of Graphs. Arxiv preprint, arXiv:1804.02473 (2018).
4. Benjamin, A., Chartrand, G., Zhang, P.: Fascinating World of Graph Theory. Princeton University Press, Princeton (2015).
5. Berliner, A.H., Dean, N., Hook, J., Marr, A., Mbirika, A., McBee, C.D.: Coprime and prime labelings of graphs. Journal of Integer Sequences, **19**(2), 1–14 (2016).
6. Bigham, A., Donovan, E.A., Pack, J., Turley, J., Wiglesworth, L.W.: Prime Labelings of Snake Graphs. The PUMP Journal of Undergraduate Research. **2**, 131–149 (2019).
7. Chartrand, G., Polimeni, A., and Zhang, P. Mathematical Proofs: A Transition to Advanced Mathematics, 4th ed. Pearson, 2018.
8. Cloys, M. and Fox, N. B. Neighborhood-Prime Labelings of Trees and Other Classes of Graphs. Arxiv preprint, arXiv:1801.01802 (2018).
9. Deretsky, T., Lee, S., Mitchem, J.: On vertex prime labelings of graphs. Proceedings of the 6th International Conference Theory and Applications of Graphs. 359–369 (1991).
10. Donovan, E., Hoard, E., Lyu, Y. and Sutton, K. On the Prime Labeling of the Disjoint Union of Cycles. Preprint (2019).
11. Gallian, J.A.: A dynamic survey of graph labeling. The Electronic Journal of Combinatorics. **16** (6). 1–219 (2009). Version 23 available at https://www.combinatorics.org/ojs/index.php/eljc/article/view/DS6/pdf.
12. "Königsberg Bridges." The MacTutor History of Mathematics Archive: http://www-history.mcs.st-and.ac.uk/history/Miscellaneous/other_links/Konigsberg.html Cited December 1, 2020.
13. Lee, J. Some Remarks on Writing Mathematical Proofs. https://sites.math.washington.edu/~lee/Writing/writing-proofs.pdf. Cited 18 December 2020.
14. Lee, S.M., Wui, I., Yeh, J.: On the Amalgamation of Prime Graphs. Bulletin of the Malaysian Mathematical Sciences Society (Second Series). **11**, 59–67 (1988).
15. Levin, O. Discrete Mathematics: An Open Introduction, $3^r d$ ed. Retrieved from: http://discrete.openmathbooks.org/dmoi3.html.
16. Patel, S. K., Shrimali, N. P.: Neighborhood-prime labeling. International Journal of Mathematics and Soft Computing. **5**(2), 135–143 (2015).
17. Rosa, A.: On certain valuations of the vertices of a graph. Theory of Graphs (Internat. Symposium, Rome, July 1966). Gordon and Breach, N. Y. and Dunod Paris, 349–355 (1967).
18. Schluchter, S.A., Schroeder, J.Z., Cokus, K., Ellingson, R., Harris, H., Rarity, E., Wilson, T: Prime labelings of generalized Petersen graphs. Involve, a Journal of Mathematics. **10**(1), 109–124 (2016).
19. Sundstrom, T. Mathematical Reasoning: Writing and Proof, Version 3. Creative Commons, 2020. Retrieved from https://scholarworks.gvsu.edu/books/24/. Cited 18 December 2020.
20. Tout, R., Dabboucy, A., Howalla, K.: Prime labeling of graphs. National Academy Science Letters-India. **5**(11), 365–368 (1982).
21. Trudeau, R. J.: Introduction to Graph Theory. Dover, New York (1994).

22. Vaidya, S.K. and Prajapati, U. M. Some new results on prime graph. Open Journal of Discrete Mathematics. **2**(3), 99–104 (2012).
23. Weisstein, Eric W."Trees." From MathWorld–A Wolfram Web Resource. Retrieved from https://mathworld.wolfram.com/topics/Trees.html. Cited 20 December 2020.
24. West, D.B.: Introduction to Graph Theory (2nd ed). Prentice Hall, Upper Saddle River (2001).

Acrobatics in a Parametric Arena

Therese Shelton, Bonnie Henderson, and Michael Gebhardt

Abstract

Parametric functions model the motion of juggling sticks, changes across time of socioeconomic factors, and more. We begin with math basics for readers who have not yet experienced the joys of parametric functions, and we move on to modeling from data sets, including some tricks with juggling "flower" sticks and competition for cell phone subscribers. Data from the flower stick and hand sticks was obtained using motion capture and digitization technologies. We note some freely and commonly available graphing tools and data sources, provide some examples and exercises, and suggest some project ideas.

Suggested Prerequisites *Ideally, readers should have successfully completed a one semester course in differential calculus, although some material is accessible to students with a solid background in pre-calculus. Polynomials, rational functions, logarithms, exponentials, and trigonometric functions are essential prerequisites.*

1 Analogies to Motivate Parametric Thinking

Suppose a lovely snowfall covers the lawn of a flat college campus on a sleepy Saturday, but the sidewalks are clear. You bundle up to go to delight in the snow, but you pause at the window and watch another person walking in the snow, sometimes going around in a square pattern, and sometimes even turning back around and

T. Shelton (✉) · B. Henderson · M. Gebhardt
Southwestern University, Georgetown, TX, USA
e-mail: shelton@southwestern.edu

retracing steps. Being the math whiz that you are, you think about the three variables involved, with an x and y pair of coordinates for each footstep, and that these were laid down in t, a time variable, according to some functions.

You head out to have some breakfast before playing in the snow, and when you return, you see a new set of tracks. You see some individual footprints, and then an oval shape of trampled snow. What you see is a static after-the-fact image that does not tell the whole story. You know that the x and y coordinates were laid down in t, again, but you cannot tell what the actual path was.

A friend sees you and calls out, asking if you have something in your room to help decorate a snow family that is being constructed. You dash to your room, and while you look for worthy contributions, you keep running to the window to take snapshots of your friend's movements.

The first example that actually has motion can be modeled with parametric functions to make an animation. We have a two-dimensional $x - y$ plane, but each of the variables x and y moves according to its dependence on a third variable t that is a *parameter*. The second example that only shows a final path might be modeled in all sorts of ways. With the right software, we can bring motion to the final path with parametric functions to implement some ideas of how may have been trod. With the third example, we can guesstimate some data from which to model the motion.

If you would prefer, think of tracks made in a flat, sandy beach instead of snow, and making sandcastles instead of a snow family.

2   Overview

We have many goals in this chapter, including having fun with mathematics, expanding the ways we think about math and its connections to the real world, thinking in multiple dimensions and from multiple perspectives, and making sense of data through analysis and modeling. Moreover, we do this especially from a minimal background in mathematics and by building on some basic tools and concepts. We hope to engage your interest and prepare you to explore some projects of your own.

To equip and inspire you, we show some work in 2D—actual, though simple, acrobatic movements of juggling sticks—and in 3D—juggling shares in a financial market. In both of these, we track multiple quantities relating to each other while changing in time. We indicate how to use some free, web-based graphing utilities, although our computations and graphs are created primarily in Mathematica and Microsoft Excel.

Working with parametric functions demonstrates the following definition of acrobatics:

> a spectacular, showy, or startling performance or demonstration involving great agility or complexity[6]

So, let us propel ourselves into the parametric arena!

3   Parametric Basics

We explore some simple introductory examples, especially for those who have never dealt with parametric functions or for anyone who could use a refresher. See the open educational resource (OER) [8] for this definition, a full development of parametric functions, and more exercises. (See 1 for other OERs that can help with the pre-requisite material.)

Definition 1 If x and y are continuous functions of t on an interval I, then the equations

$$x = x(t) \text{ and } y = y(t)$$

are called *parametric equations* in two dimensions, and t is called the *parameter*. The set of points (x, y) obtained as t varies over the interval I is called the *graph of the parametric equations,* or a *parametric curve* or *plane curve.*

3.1   Parametric Function 1

We know that the graph of

$$y = 2x + 1 \tag{1}$$

is a line. Here, y is a function of x, and x can be any real number. We can easily express this as a parametric function, just by letting $x = t$ and re-expressing y:

$$x(t) = t, \ \ y(t) = 2t + 1 \tag{2}$$

We can plot the image produced, say for $-1 \leq t \leq 2$, as we see in Fig. 1. On the left we see the static result of graphing, and on the right we add an arrow to give the direction dictated by the parameterization. Here, as t increases, we move up and to the right on the line Eq. 1.

Fig. 1 Two views of Eq. 2 for $-1 \leq t \leq 2$

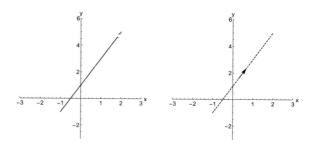

Fig. 2 A sequence of views for Eq. 2 as t increases

Fig. 3 Graph of Eq. 3 for $-1 \le t \le 2$

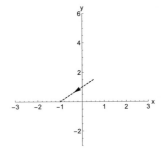

In Fig. 2, we show successive graphs for $-1 \le t \le t_{max}$ where $t_{max} = 0.2, 0.8, 1.4, 2.$, respectively. This is a way that we can indicate how an animation would look. We use a dot to mark where we begin, that is for $t_0 = -1$, and the arrow to indicate our direction at a current value of t.

Parametric functions show motion in time.

3.2 Parametric Function 2

We can change the direction on the same line in Eq. 1 by changing the parameterization.

$$x(t) = -\frac{1}{2} t, \ y(t) = -t + 1 \tag{3}$$

We can solve $x(t)$ for t, and then substitute that expression into $y(t)$ and determine that Eq. 3 also parameterizes the line Eq. 1. This process is called eliminating the parameter; sometimes it is not easy, and sometimes it is not possible.

We can plot the parametric function in Eq. 3, and even still use $-1 \le t \le 2$, just to see what happens, as in Fig. 3. If we let t run over all real numbers, we would trace out the line $y = 2x + 1$, but the new parameterization traces out just a portion of the line segment. This time, as t increases, we move down and to the left on the line.

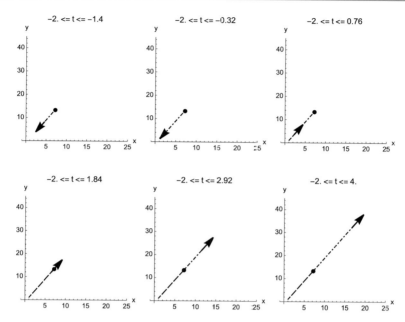

Fig. 4 Graph of Eq. 4 for different stopping points within $-2 \leq t \leq 4$

3.3   Parametric Function 3

There are innumerable ways to parameterize the line Eq. 1 to obtain all or some of the usual graph. Let us now consider

$$x(t) = 5\sqrt{t^2 + 1} - 4, \quad y(t) = 10\sqrt{t^2 + 1} - 9 \tag{4}$$

for $-2 \leq t \leq 4$.

You should verify that Eq. 4 also represents part of our line.

Each of the graphs in Fig. 4 begins at $t = -10$, near the point $(7, 13)$, but the upper value of t keeps increasing. Note that at first the parametric graph is moving down and to the left, but then it reverses course and moves up and to the right along the same line, passing the original starting point. We will soon look at Eq. 4 more closely.

4   Function Concepts

Here, we review the definition of *function* and ensure that we are aware of where the *functions* are, and how the idea is expanded, in the parametric arena.

Fig. 5 Parameterization of part of the unit circle

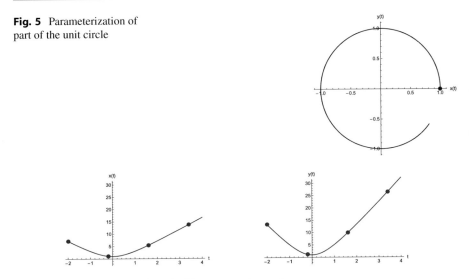

Fig. 6 Component graphs of Eq. 4 with corresponding points

4.1   Functions and Nonfunctions, Based on the Context

In our examples so far, the static 2D graph that is the end result of the parametric function, meaning with $y(t)$ graphed against $x(t)$ has represented y as a function of x as well. The graph *passes the vertical line test*, which is a visual way to think of the definition: The rule y is a *function of* x provided for each value of the *independent variable* x, the *dependent variable* y has at most one value.

With *parametric functions*, however, x and y are both *dependent variables,* and t is the *independent variable*. That is, x and y are each functions of t, but we do not always have y as a function of x in the resulting parametric graph.

A classic example of y not being a function of x in a parametric graph is given in Eq. 5 and shown in Fig. 5.

$$x(t) = cost, \ \ y(t) = sint, \ \ \ \ 0 \le t \le 0.8\pi \ \ \ \ \ \ \ \ \ \ \ \ (5)$$

Parametric graphs are much more flexible than our more usual idea of functions.

4.2   Connecting Component Functions and the Parametric Function

In each of the parametric graphs so far, the parameter t is "invisible" except in providing a direction to the graph. However, when we defined the parametric functions in Eq. 4, we defined two *component functions*. Let us graph each of these as a function of t. The component $x(t) = 5\sqrt{t^2 + 1} - 4$ is on the left in blue in Fig. 6, and $y(t) = 10\sqrt{t^2 + 1} - 9$ is on the right in red.

Table 1 Some values for
Eq. 4

t	$x(t)$	$y(t)$
−2.0	7.2	13.4
−0.2	1.1	1.2
1.6	5.4	9.9
3.4	13.7	26.4

Fig. 7 Component Graphs of
Eq. 4 connecting
corresponding points for
$t = -2$

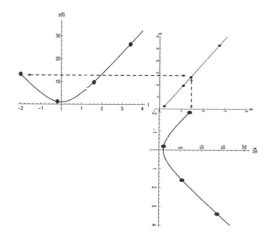

Points are indicated for $t = -2.0, -0.2, 1.6, 3.4$. These are tabulated in Table 1 and displayed prominently in Fig. 6. Such a table can help us know the direction for a parametric curve, and more.

Now we relate the separate graphs in Fig. 6 to a parametric view. We rotate the blue $x(t)$ graph from the left of Fig. 6 and align its x-axis underneath the horizontal axis for the parametric graph. We show the red $y(t)$ graph from the right of Fig. 6 to the left of the parametric graph, and we align the vertical axis of each. See the way we have "tumbled" our graphs from Fig. 6 with the parametric graph, shown in purple in Fig. 7. We also connect points corresponding to $t = -2$.

In Fig. 8, we connect more of the points in Table 1.

In Figs. 6 and 7, we can see that at least for some negative values of t, as t increases, $x(t)$ decreases, and $y(t)$ also decreases. For these values of t, the purple parametric graph in Fig. 7 has movement down and to the left. At least for some positive values of t, as t increases, $x(t)$ increases, and $y(t)$ also increases. This is when the purple parametric graph has movement up and to the right. When does it change behavior? There are multiple approaches we can take when answering this question: guesstimate values from the component graphs, watch dynamic graphs, build extensive tables of values, and we can use calculus.

For Eq. 4, we have

$$\frac{dx}{dt} = \frac{5t}{\sqrt{1+t^2}}, \quad \frac{dy}{dt} = \frac{10t}{\sqrt{1+t^2}} \tag{6}$$

Fig. 8 Component Graphs of
Eq. 4 with more
corresponding points

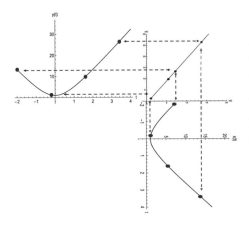

We see that $\frac{dx}{dt}$ is 0 for $t = 0$, negative for $t < 0$, and positive for $t > 0$. Therefore, we see in Fig. 6 that $x(t)$ decreases for $t < 0$ to its minimum value at $t = 0$, and then increases for $t > 0$. We see a similar behavior for $\frac{dy}{dt}$. Combining these, the parametric curve moves down and to the left for $t < 0$, and then up and to the right for $t > 0$.

Challenge Problem 1 What types of component functions lead to a parametric function which also represents y as a function of x? What *behaviors* or characteristics are needed? How can we determine those from component graphs? How can we use calculus to identify possible intervals of the parameter for which this is true?

5   Try Some Parametric Acrobatics Yourself

Now you can try a greater variety of parametric functions, including other parameterizations of the line we graphed previously. In this section we explore a freely available online graphing tool that can be used to explore a rich variety of other parametric functions, too; included exercises should help interested readers become more familiar with both this computational tool and the underlying concepts.

5.1   General Desmos Graphing Instructions

We use a web-based tool that is, at the time of this writing, free and freely available: https://www.desmos.com/calculator. Users may take some screenshots or download a graph, and users may create a free account to save their graphs.

For any parametric graphing using the form

$$x = f(t), y = g(t)$$

input the functions as an ordered pair into the first cell.

$$(f(t), g(t))$$

Under the ordered pair, set real number boundaries for t,

$$a \le t \le b$$

With this input you will get a segment of the graph that is between $t = a$ and $t = b$
In the next cell, input the same ordered pair except swap t for a

$$(x, y) = (f(a), g(a))$$

Under the ordered pair, there will be an option to add a slider for a. Click this button and use the same boundaries for a as you did t This will give you a point that will trace the graph segment you have already graphed
On the left of the second cell, push the play button and see the point trace the line. Under the play/pause button, click the button with "loop" arrows to access animation properties. Change the animation mode to the second option, with the two arrows pointing the same direction. This makes it so the point repeats going in the same direction (goes back to the beginning of interval after it reaches the upper boundary). Indicate the endpoints and the step size for the animation.
Desmos also has tutorials, such as https://learn.desmos.com/parametric-equations.

5.2   Exercises with some Desmos Instructions

1. Let $x = 4t + 6$ and $y = 5t + 1$. Graph the parametric function from $t = -1$ to $t = 1$ and create a point that traces a recognizable curve.

   - In the first cell input the ordered pair $(4t + 6, 5t + 1)$
   - Set the interval to $1 \le t \le -1$
   - Adjust the window of the graph by pinching and dragging so that the full line segment is in view
   - For the point, go to the next cell and input $(4a + 6, 5a + 1)$
   - Add a slider for a, using the interval $-1 \le a \le 1$
   - Press the play icon and change the animation option "repeat in same direction"
   - Observe the direction and speed of the point as it moves along the graph

2. Let $x = t + 1$ and $y = 2t^2$. Graph the parametric function from $t = -1$ to $t = 1$ and create a point that traces a recognizable curve.

- In the first cell input the ordered pair $(t + 1, 2t^2)$
- Set the interval to $1 \leq t \leq -1$
- Adjust the window of the graph by pinching and dragging so that the full segment of the parabola is in view
- For the point, go to the next cell and input $(a + 1, 2a^2)$
- Add a slider for a, using the interval $-1 \leq a \leq 1$
- Press the play icon and change the animation option "repeat in same direction"
- Observe the direction and speed of the point as it moves along the graph. When does the point speed up or slow down? Why does it change speed? What does it say about the function around the points at which it changes speed?

3. Using the same step as above, graph $x = 5\,cos(t)$, $y = 2\,sin(t)$ on the interval $t = 0$ to $t = 2\pi$. Plot a point that traces a curve, and note where the point changes speed.
4. Using the same step as above, graph $x = sec(t)$, $y = 3tan(t)$ on the interval $t = 0$ to $t = 2\pi$. Plot a point that traces a curve, and note where the point changes speed.

5.3 Experimentation

This problem will probably require multiple tools, one of which may be Desmos.

Challenge Problem 2 Analyze our classic example, Eq. 5 as we did in Sect. 4.2. Create a table of values, graph the component functions, and relate them to the parametric function. What other types of parameterizations of the unit circle could there be? Experiment and state some. Write about your findings.

6 True Acrobatics: Parametric Modeling of a Flower Stick

Now we combine the physical tumbling of a flower stick with the mental acrobatics of parametric functions.

6.1 Flower Sticks and Digitized Motion

Flower sticks, also known as devil sticks (except for minor design differences), are sticks that are, in a way, juggled. While traditional juggling includes keeping three balls off the ground while only one can ever be in the hand, a flower stick is kept in the air using two other sticks that are held in the right and left hands, which we call hand sticks. ("Flower Sticks" can be used to describe all three sticks, but we try

Fig. 9 Flower stick trick presentation

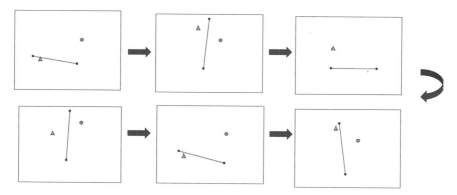

Fig. 10 Processed flower stick and two hand sticks

to be clear.) The hand sticks then rotate the flower stick, and this motion keeps the flower stick from hitting the ground.

We consider a basic trick known by the following names: a standard idle, a ticktock, or single sticking. In this trick, the flower stick is simply knocked back and forth between the two hand sticks, and there are no extra turns or movements. Author B. Henderson performed and analyzed this and other tricks while she was an undergraduate student, supervised by author T. Shelton.

Figure 9 shows still shots taken from a video of the trick during an on-campus presentation of the mathematical modeling.

In Fig. 10 we see a sequence of some of the processed digitized data. The line connected by the red and blue dots is the flower stick, and the orange triangle and green circle are the ends of the sticks in the left and right hands, respectively. The video[1] is also available.

To obtain these digitized images, reflective markers were placed on both ends of the flower stick as well as the ends of the two hand sticks closest to the video camera. The performer was positioned against a wall, on which there were markers to determine distance. An industrial light source was then directed at the sticks and the performer. This light source highlighted the reflective markers, which allowed the video camera and software to more easily recognize the markers. The resulting spreadsheet data included the frame number, and the x and y coordinates of each

Fig. 11 Digitization of flower stick trick

Fig. 12 x-coordinate in time of left end of the flower stick

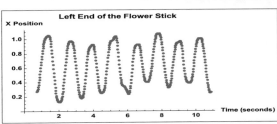

marker. Dr. McLean conducted the digitization using equipment and software from the Department of Kinesiology for analyzing movement. Figure 11 shows frames from the video capture of the digitization process; the reflectors are clearly seen. The low-resolution of these images was adequate for the motion analysis software.

Some minor modifications to the images were made in order to more clearly highlight the trick being performed; a few initial frames were deleted and "frame numbers" were converted to "time in seconds" based on a sampling rate of fifty frames per second. Now we begin the analysis.

6.2 Modeling the Left End of the Flower Stick

The data for the x coordinate, shown in Fig. 12 looks sinusoidal. There are a number of somewhat advanced techniques for fitting sines and cosines to data, which were beyond the scope of this student modeling project. (The interested reader might read about "Fourier Analysis.") We simply estimated the period from the data, built a list of sines and cosines based on this period, and then used Mathematica to determine coefficients for a sinusoidal representation of the x position with respect to time t of the left end of the flower stick. We omit detail of the modeling to maintain the focus on the resulting parametric function,

$$x_L(t) = -1460.00 sin(16.30t) + 364.21 sin(17.08t) +$$

$$6.00 sin(17.48t) + \ldots + 115.09 cos(17.082t) + 21.81 cos(17.48t)$$

(omitting many terms.) From a pure modeling standpoint, we would balance the complexity of the model with the closeness of the fit, but for this undergraduate research project, the student was encouraged to explore freely.

Fig. 13 x-coordinate in time of left end of the flower stick

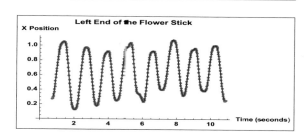

Fig. 14 Left end of the flower stick

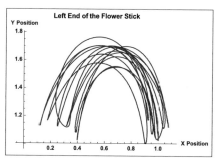

The resulting function $x_L(t)$ fits the data well for the x coordinate of the left end of the flower stick, as we see in Fig. 13 (The modeling function is shown in red, and the original data is blue).

Similarly, we fit a sinusoidal function $y_L(t)$ to the y coordinate of the left end of the flower stick.

Now for the exciting part—we now have a parametric function $(x_L(t), y_L(t))$ for the left end of the flower stick, which we graph in Fig. 14.

Looking back at Fig. 10, we can tell that the left end of the flower stick does, indeed, travel back and forth across the top of our viewing window. So, what should that mean for the right end of the flower stick? Think about this while you view Fig. 15, the component and parametric graphs together for the left end of the flower stick.

Notice that in Fig. 14, y is not a function of x at all, and it is only by fitting curves to the component functions that we are really able to model the full movement. The view in Fig. 14 is the cumulative movement shown all together, like the analogy of walking in the snow. We might actually see something like this if we were in the dark, watching someone juggle a flower stick that had a small light source on the left end, when our eyes continue to see some light after it has passed a point.

Using descriptive details like this can help us solidify the connections between a model and reality.

6.3 More on the Flower Stick

Similarly, we fit functions $x_R(t)$ and $y_R(t)$ for the x and y coordinates of the right end of the flower stick, respectively.

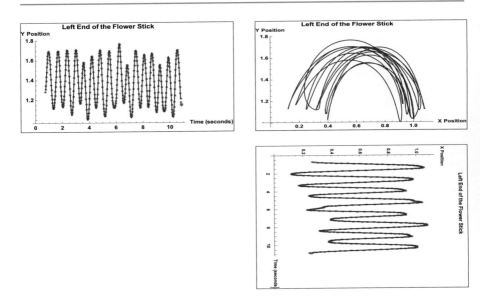

Fig. 15 Left end component and parametric views

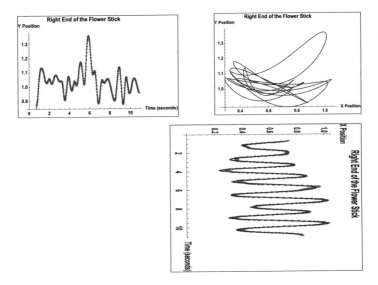

Fig. 16 Right end component and parametric views

From Fig. 16 it seems that there is a single cycle that was much more wobbly in height than the other cycles, although the width of that wobble is consistent with the widths of the other cycles. We can see the wobble both in the $y_R(t)$ graph and the parametric plot, and we can see the greater consistency in the widths both in the

$x_R(t)$ graph and the parametric plot. We encourage the reader to use the component graph(s) to identify the time interval over which this wobble occurred.

We can also compare the movement of the left end and the right end. The right end moves less consistently than the left end, although we note that the scales are not exactly the same. This seems to make sense by examining the position of the hand sticks, relative to the flower stick, in Fig. 10. The left end of the flower stick is being batted back and forth by the two hand sticks, restricting its movements more, while the right end can swing more.

6.4   Your Work

Challenge Problem 3   Analyze our classic example, Eq. 5 as we did in Sect. 4.2. Create a table of values, graph the component functions, and relate them to the parametric function. What other types of parameterizations of the unit circle could there be? Experiment, and write about your findings.

Challenge Problem 4   Perhaps you have personal data, such as from a device that monitors your movements, heart rate, sleep modes, etc. that you could analyze. You need two sets of data linked to the same times.

Is your data periodic? if so, estimate the period, perhaps from peak to peak.

Where is the data increasing? decreasing? Does it have extrema?

Estimate the velocity.

Examine concavity and any changes, indicating where velocity increases or decreases.

Interpret the functions and what they mean about your movements or health.

Depending upon how extensive your data and analysis are, this could develop into more of a Research Project.

Research Project 1   Access and download the data for the digitization of the Flower Stick trick from [1]. This includes the frame number, x and y coordinate data for each end of the Flower Stick, and x and y coordinate data for each hand stick. Perform your own analysis.

You might want to begin with the smaller data set provided.

Create component and parametric graphs for appropriate combinations of the data. Can you use spreadsheet functions, graphs, or computations to identify the period and amplitude for sinusoidal data?

Compute the average rate of change within a variable, and use that to help identify local extrema of the variable, as well as intervals of increase or decrease of the x and y components.

(continued)

Pick an interval that has at least one full period for related x and y components, and formulate a model for that time-section of the data. Can you model that section? Graph your parameterization of that section for t values beyond that time-section. If you repeat it, can you mimic the trick just with this model?

Experiment, and write about your findings.

7   Financial Acrobatics: Modeling US Wireless Subscribers

Increasing consumer reliance on cell phones is apparent in Table 2, as is increased pressure on companies to construct new and innovative ideas. By predicting future consumer numbers, a company is better prepared to set priorities and project the number of people they might serve. We created models in 2015 for ATT, Verizon, and T-Mobile from 2004–2014 in the USA. We show some of these, and compare with a bit more current information. As with the Flower Sticks project, the focus at the time was on mathematical modeling, details of which are omitted here. Author M. Gebhardt gathered and analyzed the data while he was an undergraduate student, supervised by author T. Shelton.

Table 2 US thirst for cell phones

Year	t	Subscriptions per 100 people	Population (millions)	Subscriptions (millions)
2004	0	63.22	2928.05	185.10
2005	1	69.05	2955.17	204.06
2006	2	77.11	2983.80	230.08
2007	3	82.93	3012.31	249.82
2008	4	86.10	3040.94	261.82
2009	5	89.55	3067.72	274.70
2010	6	92.27	3093.22	285.40
2011	7	95.45	3115.57	297.38
2012	8	97.07	3138.31	304.63
2013	9	98.20	3159.94	310.30
2014	10	111.56	3183.01	355.08
2015	11	119.14	3206.35	382.02
2016	12	122.56	3229.41	395.79
2017	13	123.11	3249.86	400.08
2018	14	123.69	3266.88	404.07
2019	15	123.69	3282.40	405.99

7.1   About the Data

Some data came from online annual reports; some were recently obtained from [2] and [7]. Data from [9] on overall population and density of cellphone subscriptions was used in the original model and the current comparisons.

We simplify descriptions of some of the data, for instance, by using "cell phones," "mobile devices," "subscribers," "wireless subscriptions," and "connections" inter-changeably. We also ignore various distinctions, such as between "cell phone" and "smart phone," between "pre-paid" and "post-paid," etc. Variations in the "who, what, how, when, and why" of data collection and reporting leads to multiple variations in data sets. Many of us have been made more aware of this phenomenon from public reporting, such as on the COVID-19 pandemic: data was posted for "cases" of those infected with the coronavirus, but this may or may not have included "suspected cases," and often there was a backlog of data. Often when we seek data, as is the case with the cell phone data, we find that the latest is from years ago, and we often need to cobble together data from multiple sources.

7.2   Views of the Data

These three companies had a large market share in 2015, encompassing most cell phone users in the USA. Data is listed in Table 3, measured in millions of subscriptions.

Both ATT and Verizon led the US cellphone market in wireless subscribers by quite a large margin over T-Mobile. However, T-Mobile achieved a dramatic increase from 2012 to 2013 from beginning to sell the iPhone and its acquisition of MetroPCS. Improvements to T-Mobile's network and customer experience contributed to the first year of increases and also accounted for the second year of increases revealed in the data.

We did not know in 2015 whether any of these companies would persist, nor did we know that T-Mobile would acquire Sprint, a merger that was finalized in

Table 3 Cell phone subscription data, in millions

Year	t	AT&T	Verizon	T-Mobile
2004	0	49.132	43.800	1.399
2005	1	52.144	51.300	1.925
2006	2	60.962	59.100	2.941
2007	3	70.052	65.700	3.963
2008	4	77.009	72.100	5.367
2009	0	85.120	85.445	6.640
2010	1	95.536	87.535	8.155
2011	2	103.247	92.167	9.347
2012	3	106.957	96.700	8.887
2013	4	110.400	102.600	43.000
2014	5	120.554	108.200	55.000

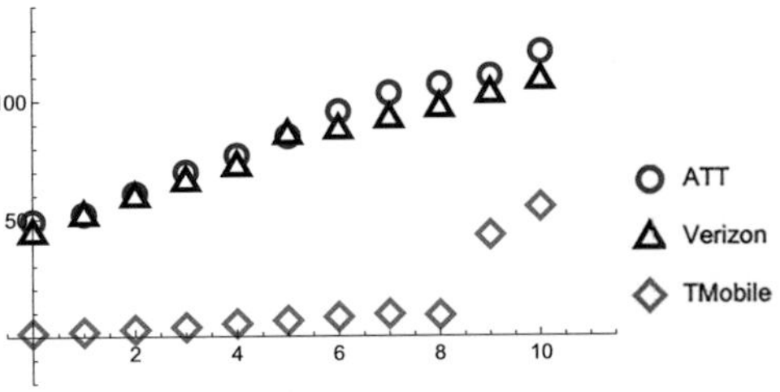

Fig. 17 Cell phone
subscription data, in millions,
vs years since 2004

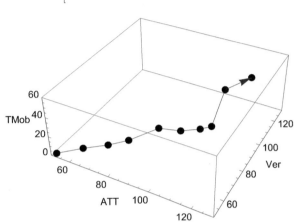

Fig. 18 Parametric view of
cell phone subscription data,
in millions with an axis per
company

April 2020 [3]. Sprint was also a key player, and for many years was comparable to
T-Mobile. We choose to focus on the three main companies.

The data from Table 3 for each company across time is visualized in Figs. 17
and 18. Figure 17 shows each company's subscriber data separately as variables
dependent upon t, years since 2004.

The data is perfect to be displayed as a parametric graph because the parameter,
years since 2004, is the common parameter for the sets of subscriber data for the
companies. Here, we see a red arrow at the end to indicate increasing time t. The
view in in Fig. 18 focuses on relating the companies to each other. Compare the
scale for each company. T-Mobile stays on the "floor" until the last two years, and
even then does not even rise to half of the number of subscribers for either AT&T
or Verizon.

7.3   Data Projections

We briefly outline some of the models we created based on this data, meaning that
we leave out most of the detail of the modeling used. Our purpose is to encourage
experimentation and critical thinking. A spreadsheet can be implemented for at
least parts of most of these models. We show five-year projections of the modeled
equations to 2019, recalling that extrapolating beyond the data is always risky. We
include updated data from Table 4 in the graphs to evalluate our models.

Table 4 Updates to cell phone subscription data, in millions

Year	t	AT&T	Verizon	T-Mobile
2015	11	128.68	140.92	63.28
2016	12	134.88	145.86	71.46
2017	13	141.20	150.46	72.59

Our mathematical *models* produce equations using the initial data from 2004 through 2014, which we graph through 2019. Rather than modeling with the additional data from Table 4 through 2017, we demonstrate a methodology of conjecture and validation. Moreover, we leave the task of modeling all (or most) of the data to you. Our models may give you some ideas for your own analysis.

Research Project 2 Create component and parametric graphs for appropriate combinations of the cell phone data. You may wish to find further updates to the data, as well as to withhold the last year to validate your results. If the data seems to change dramatically, you may wish to ignore some of the initial years, or break the data into parts and find fits to the subsets.

Use spreadsheet functions, graphs, and computations to formulate models.

You may wish to compute rates of change within a variable, and use that to help identify local extrema of the variable, as well as intervals of increase or decrease of the x and y components.

Graph your parameterizations for t values beyond the data, though not too far beyond.

Experiment, and write about your findings.

7.4   Linear Fits to 2004–2014 Data

The 2004–2014 data for AT&T and Verizon appear very linear, and data for T-Mobile also appears linear through 2011. We see in Fig. 19 that the functions in Eq. 7 appear to fit reasonably well, although the linear model clearly does not capture the dramatic rises for T-Mobile.

$$A(t) = 47.69 + 7.40t,$$

$$V(t) = 46.417 + 6.44t, \tag{7}$$

$$T(t) = -7.75 + 4.22t$$

For these linear fits to the 2004–2014 data, the coefficients of determination, R^2 values, are 0.99, 0.98, and 0.66, respectively. These may only be interpreted meaningfully when a graph verifies that the data is reasonably linear. The values close to 1 for AT&T and Verizon verify the good fit, with almost 100% of the

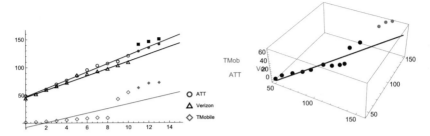

Fig. 19 Data with linear model in component and parametric graphs

Table 5 Annual proportional growth	t	AT&T	Verizon	T-Mobile
	1	1.06	1.17	1.38
	2	1.17	1.15	1.53
	3	1.15	1.11	1.35
	4	1.10	1.10	1.35
	5	1.11	1.19	1.24
	6	1.12	1.02	1.23
	7	1.08	1.05	1.15
	8	1.04	1.05	0.95
	9	1.03	1.06	4.84
	10	1.09	1.05	1.28
	Average	1.10	1.10	1.63

variability in the cell phone data accounted for by the model. The significantly lower value for T-Mobile verifies the looser fit of the linear model, with only 66% of the variability in the cell phone data accounted for by the model.

7.5 Proportional Growth

We see a pattern in the ratios of the number of subscribers in a subsequent year to the number of subscribers in the current year, although the pattern breaks down for T-Mobile.

The data from Table 5 suggests that we consider an exponential fit.

Challenge Problem 5 Can you explain, or compute, a relationship between proportional growth and exponential functions?

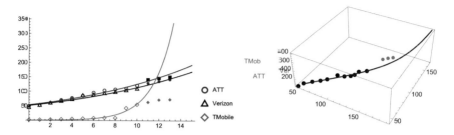

Fig. 20 Data with exponential model in component and parametric graphs

$$A(t) = 53.43 \; e^{0.085t},$$

$$V(t) = 51.37 \; e^{0.079t}, \tag{8}$$

$$T(t) = 0.28 \; e^{0.053t}$$

We have slightly improved fits for AT&T and Verizon, and a significantly improved fit for T-Mobile, as seen in Fig. 20. For these exponential fits to the data through 2014, the coefficients of determination are 0.998, 0.997, 0.952, respectively. However, the extra data reveal that T-Mobile's improvements do not continue in the same manner as the extraordinary two years, which is reasonable. However, perhaps such a model seems to "predict" the merger with Sprint.

8   GeoGebra to Practice 3D Parametric Equation

Here are a very brief set of instructions to get you started on a free graphing tool, along with some problems to build, or examine, your ideas about parametric functions.

- Go to the following parametric graphing website: https://www.geogebra.org/3d?lang=en
- Label your x-axis, y-axis, and z-axis with the settings button in the top right
- You will be graphing a series of 3D functions. For the following examples, type $curve(x(t), y(t), z(t), t, a, b)$ into the input slot, where $x(t)$, $y(t)$, and $z(t)$ are components of the parametric function, and t is on the interval $[a, b]$

The following exercises can be used to build your understanding of GeoGebra and 3D Parametric Equations.

1. Set up the following functions on the interval $[-20, 20]$

$$x(t) = t^2$$
$$y(t) = t$$
$$z(t) = t$$

   a. Use your mouse to move the graph around, notice how it looks from different angles. What 2D graph does it resemble in each plane?

2. Now try these on the interval $[0, 2\pi]$

$$x(t) = 5\,cos(t)$$
$$y(t) = 5\,sin(t)$$
$$z(t) = 0$$

   a. Notice the graph is only in 2 dimensions, why is this the case?
   b. Try changing $z(t)$ to a different number, what happens to the graph?
   c. Now change it to $z(t) = t$, how did the graph change? Follow the flow of the graph starting at $t = 0$, notice how each equation's value matches up with the graph's location in each axis. Try increasing the upper bound to see the graph travel all the way up the z-axis

3. Still on the Interval $[0, 2\pi]$

$$x(t) = t$$
$$y(t) = 5\,cos(t)$$
$$z(t) = 5\,sin(t)$$

   a. Notice we only switched the equations around from example 2. How did this change the graph? In what plane(s) is the new graph symmetric?

4. Lastly, on $[0, 6\pi]$. How do you think this graph will look, using examples 2 and 3 as reference?

$$x(t) = t\,cos(t)$$
$$y(t) = t\,sin(t)$$
$$z(t) = t$$

   a. Explain the shape of the graph.

9   Your Project

There is a rich variety of topics to explore and model with parametric functions, and you are encouraged to explore them!

Research Project 3 Come up with your own project. Some ideas:

1. your own motion capture data, if you have access to some technology
2. pollen count data aligned with your own record of numerical scores to indicate the severity of your allergy symptoms
3. tracking social justice data over time, such as comparing proportions of subgroups in a population to proportions of those subgroups who receive (or do not receive) services or retributions. This might include data from the Covid-19 pandemic, perhaps from Johns Hopkins[5].

GapMinder[4] has marvelous animations of historical data shown in 2D, and the data is downloadable. Exploring this site can spur numerous ideas.

We hope you will launch into some parametric investigations to learn more about mathematics and about your world.

Acknowledgments We gratefully acknowledge the following:

Research Assistants E. Wilson Cook and Audrey Schumacher, for instructions for Desmos and GeoGebra, as well as some fun examples. Research Assistant Emily Thompson for mathematical typesetting and some editing.

Southwestern University, for funding to support the work of the Research Assistants.

Dr. Scott McLean, Kinesiology Department, Southwestern University for working with author B. Henderson and obtaining data from her Flower Stick tricks.

References

1. Author files available for download for this chapter. https://drive.google.com/drive/folders/197a-lWnnDk-7DNihKTGpakqT373uP8vj?usp=sharing
2. Annual Reports of cellphone companies, https://www.annualreports.com/. Cited 30 Jan 2021.
3. Conrad, Roger. "T-Mobile US: The Sprint Merger So Far." Forbes.com. 23 Jun 2020. https://www.forbes.com/sites/greatspeculations/2020/06/23/t-mobile-us-the-sprint-merger-so-far. Cited 30 Jan 2021.
4. GapMinder, from the Swedish nonprofit foundation Stiftelsen Gapminder. https://www.gapminder.org/. Cited 30 Jan 2021.
5. Johns Hopkins Coronavirus Resource Center. https://coronavirus.jhu.edu/. Cited 28 Jun 2021.
6. Merriam-Webster, Incorporated, https://www.merriam-webster.com/dictionary. Cited 30 Jan 2021.
7. "Mobile Wireless Competition Reports (19th Annual)" United States Federal Communications Commission. Publish Date: 23 Sep, 2016 https://www.fcc.gov/reports-research/reports/mobile-wireless-competition-reports. Cited 30 Jan 2021.
8. Strang, G., et al. OpenStax Calculus, Volume 2, Chapter 7. Publish Date: 30 Mar, 2016. Web Version Last Updated: 21 Dec, 2020. https://openstax.org/details/books/calculus-volume-2 Cited 30 Jan 2021.
9 World Bank Group. https://data.worldbank.org/ Cited 30 Jan 2021.

Further Reading

1. OpenStax Math, open resource books that include pre-calculus. https://openstax.org/subjects/math
2. Stover, Christopher and Weisstein, Eric W. "Parametric Equations." From MathWorld–A Wolfram Web Resource. https://mathworld.wolfram.com/ParametricEquations.html.

Software

1. Contemplas Templo, a motion analysis software platform, and Vicon Motus, a 2D video-based motion capture and analysis tool, both developed by CONTEMPLAS GmbH.
2. Desmos
3. GapMinder
4. GeoGebra
5. Mathematica, by Wolfram, was used for other analysis and graphs.

But Who *Should* Have Won? Simulating Outcomes of Judging Protocols and Ranking Systems

Ann W. Clifton and Allison L. Lewis

Abstract

We all know the world loves an underdog. But who *should* have won? Simulation models can give us a way to reproduce contest scenarios using computer power, allowing us to simulate versions of reality numerous times and observe a wide range of possible outcomes. Using basic probability calculations and a wide range of programming tools, we can mimic real life and answer such questions as "how likely is Team A to win the tournament?", "how biased would Judge B have to be in favor of an athlete for them to come out on top?", or "is there a particular voting structure that would allow for the emergence of Candidate C as the winner?" In this chapter, we lay out the fundamental skills for pursuing answers to questions like these via simulation, and leave the interested reader with a set of potential research projects on the theme of evaluating fairness, bias, and other issues relevant to ranking schemes and judging protocols.

Suggested Prerequisites *A basic familiarity with programming software of choice (R, Python, MATLAB, Java, C++, or others). Introductory courses in probability or computer programming may be helpful, but are not required.*

A. W. Clifton
Louisiana Tech University, Ruston, LA, USA
e-mail: aclifton@latech.edu

A. L. Lewis (✉)
Lafayette College, Easton, PA, USA
e-mail: lewisall@lafayette.edu

1   Introduction

One of the most powerful tools in an applied mathematician's or statistician's toolbox is simulation. What is simulation? Simulation is a method by which we can exploit computer power to mimic some real-world behavior or data collection indirectly. Furthermore, simulation allows us to generate huge numbers of samples for analysis without the cost of performing the experiment manually. For example, perhaps I want to count how many heads I'll get if I flip a coin 1000 times. Rather than spending my whole afternoon flipping and recording, I can write some quick code in my favorite programming language and have the computer simulate 1000 flips of a coin in under a second, using just a few simple assumptions! Of course, we can simulate far more complex scenarios than just coin flips—all we need are a few basic tools.

Simulation also gives us the ability to analyze a scenario under a variety of different conditions. In many cases, running an experiment multiple times in the real world would be prohibitively expensive or even completely infeasible, but by building a computer simulation that mimics that real-world experiment, we can still gain insight into how the scenario might play out in reality. We can tweak key parameters or change underlying assumptions, and make observations about what kinds of impacts those perturbations would have on the system. Simulation is a particularly powerful tool in support of the *frequentist perspective of probability*. Frequentists define probability using the concept of relative frequency; that is, the probability that an event will occur is measured by repeating an experiment many times and observing the long-term frequency of that outcome. Thus, to use the frequentist perspective, an experiment must be repeatable over a great many number of trials. Even if the experiment can't be easily repeated in real life, computer simulation can still give us a way to estimate these probabilities.

Our initial goal for this chapter is to teach students the basic rules of probability, to assist them in solving simple problems, and to help them develop intuition for more complex scenarios. This foundation will be provided in Sect. 2, but can also be gained from nearly any introductory statistics text which includes a probability chapter—we specifically recommend [1, 6, 7, 11] for the beginning probability student, and [2, 4] for more advanced students seeking a calculus-based approach to probability. We'll then proceed to the fundamentals of simulation via computer programming in Sect. 3, allowing students to further explore more complicated models. Throughout the chapter, we'll build up to an investigation of relevant applications on the theme of contests. Who wins, and under what conditions? What happens if a judge is biased? Can two different ranking schemes result in two different winners? How do you decide which scheme is fair? Section 4 lays out several potential research projects on these themes, where students will be prompted to explore a range of scenarios through the development of simulation code.

Students tackling this material should have access to programming software such as R, Python, MATLAB, Java, or C++. Throughout this chapter, we'll consistently provide sample code for—and use the notation of—the open-access statistical software, R, though students can easily adapt our provided sample code to another program of their choice. Students opting to use R can download the software at https://www.r-project.org, or access it through a web browser at https://rstudio. cloud. Though we will teach all necessary simulation tools within the chapter itself, we do assume students will have basic familiarity with the setup and key features of their chosen language (i.e., how to start a new script, define variables, use basic calculator functions, etc.). For students looking for an additional resources for getting started in R, we recommend [5, 12].

2   Fundamentals of Probability

Probability is the study of chance, randomness, and uncertainty. If you've ever responded to a question by saying "Probably!" or "Most likely!", then you've already been using the language of probability! The study of probability began with games of chance. The rolling of a standard six-sided die is a common example—in the language of probability, we can refer to a single die roll as an *experiment*. In this setting, an experiment is simply a process whose outcome is uncertain. Although the outcome of a die roll is unknown prior to rolling, we do know that there are only six possibilities that can occur. We often refer to this set of possible outcomes as the *sample space*. Since this set lies at the foundation of all of our subsequent probability calculations, we will begin our discussion with some relevant ideas from set theory.

2.1   A Little Set Theory

When you roll a standard die, there are six possible outcomes. When you flip a coin, there are two possible outcomes. When creating your course schedule, there may be too many possible outcomes to easily count! In each of these cases, we think of the possible outcomes as a *set* of objects. We can represent the set of outcomes— or sample space—when rolling a standard six-sided die as $S = \{1, 2, 3, 4, 5, 6\}$. Similarly, we could represent the set of outcomes for flipping a coin as $S = \{H, T\}$, where H stands for heads and T stands for tails. For some other examples, such as designing a course schedule, the full set of possible outcomes is much more difficult to list.

Each distinct object in a set is called an *element*. If a set does not contain any elements, it is called the empty set and denoted $\emptyset$. A collection of elements of interest is often referred to as an *event*. For instance, the event of obtaining an odd number on a die roll could be denoted by $A = \{1, 3, 5\}$. An event, A, which is entirely contained within another event B is said to be a *subset* of B, written $A \subset B$. For example, we could say that the event $A =$ obtaining an odd number when rolling a die is a subset of event $B =$ obtaining a prime number when rolling a die, since $\{1, 3, 5\} \subset \{1, 2, 3, 5\}$. To allow for the possibility that the sets A and B are the same, we write $A \subseteq B$. Two sets A and B are said to be *equal*, $A = B$, if and only if $A \subseteq B$ and $B \subseteq A$.

We're often interested in exploring what happens when we combine sets that represent different events. For instance, how likely are we to get a number that is both odd AND prime? How about a number that is odd OR a multiple of three? This leads us to the following definitions:

Definition 1 The *union of two sets* A_1 and A_2 is the set that contains all outcomes that belong to *at least one of* the two sets. It is denoted

$$A_1 \cup A_2.$$

For example, in our die rolling experiment, if

$$A = \text{rolling an odd number} = \{1, 3, 5\}$$

and

$$C = \text{rolling a multiple of three} = \{3, 6\},$$

we would conclude that $A \cup C = \{1, 3, 5, 6\}$. That is, we could observe the event $A \cup C$—the event of rolling an odd number OR a multiple of 3 OR both—by rolling any of the numbers 1, 3, 5, or 6. Note that even though the number 3 is contained in both sets A and C, it appears in the union set just once.

Definition 2 The *intersection of two sets* A_1 and A_2 is the set that contains the outcomes *common to both of* the sets. It is denoted

$$A_1 \cap A_2.$$

For example, in our die rolling experiment, if we return to

$$A = \text{rolling an odd number} = \{1, 3, 5\}$$

and

$$B = \text{rolling a prime number} = \{1, 2, 3, 5\},$$

then $A \cap B = \{1, 3, 5\}$. This is because only the numbers 1, 3, and 5 are contained in *both* sets. In other words, the intersection represents the set of all possible outcomes that would satisfy events A and B simultaneously.

The definitions of union and intersection can be extended to n sets where n is any natural number. We use the following notation:

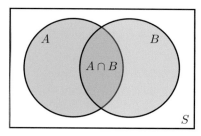 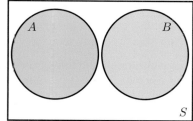

Fig. 1 Simple Venn diagrams showing relationship between two events A and B. On left, two events A and B share some outcomes. On right, A and B are mutually exclusive events

$$\bigcup_{k=1}^{n} A_k = A_1 \cup A_2 \cup \cdots \cup A_n = \text{elements in } any \text{ set } A_1,\ A_2, \ldots, A_n$$

and

$$\bigcap_{k=1}^{n} A_k = A_1 \cap A_2 \cap \cdots \cap A_n = \text{elements in } all \text{ sets } A_1,\ A_2, \ldots, A_n$$

to denote the union and intersection of n sets, respectively.

It often helps to have a way to visualize the overlap between different events. We can use Venn diagrams to achieve this. In Fig. 1, we show a standard two-event Venn diagram. Events A and B are drawn as subsets of the sample space S, with their overlap labeled as $A \cap B$.

Exercise 1 For each of the following, shade in the corresponding area on a standard two-event Venn diagram:

a) $A \cup B$ b) A^c c) $A^c \cup B$

d) $A^c \cap B$ e) $(A \cup B)^c$ f) $(A \cap B)^c$

Exercise 2 Let A be the event that you roll an odd number, B represent the event that you roll a multiple of 3, and C be the event that you roll a prime number in a single die roll. Show that the following properties hold:

a) $A \cup (B \cup C) = (A \cup B) \cup C$ b) $A \cap (B \cap C) = (A \cap B) \cap C$

We call these properties the *associative properties*, and they hold for all sets A, B, and C.

What happens if two sets have no elements in common? Sets with no overlapping elements play a large role in the study of probability, and merit their own set of definitions.

Definition 3 Two sets A and B are said to be *disjoint* or *mutually exclusive* if they do not share any common outcomes. That is, A and B are disjoint if $A \cap B = \emptyset$, where $\emptyset$ denotes the empty set.

For example, in our die rolling scenario, we would say that the sets $P = $ rolling a number less than 3 and $Q = $ rolling a number greater than 4 are disjoint events, since $P = \{1, 2\}$ and $Q = \{5, 6\}$ do not share any of the same outcomes.

Definition 4 Two sets are said to be *complements, or complementary*, if they are disjoint and cover the entire sample space, S; that is, $A \cap A^c = \emptyset$ and $A \cup A^c = S$, where A^c denotes the complement of A.

Exercise 3 Using the example of a die roll, construct two sets of outcomes that are complementary.

Exercise 4 Is it possible for two sets A and B for which $A \subseteq B$ to be disjoint? Either construct an example in which this holds or explain why such a statement cannot be true.

2.2   Computing Probability

Now that we've developed a foundation in working with sets and events, we're ready to discuss how we compute the probabilities with which certain events can be expected to occur. Recall, we consider an event to be a subset of possible outcomes of an experiment. We denote the number of outcomes in event E by $|E|$.

When an experiment contains $n \geq 1$ possible outcomes for some natural number n, and all outcomes are equally likely, we compute the probability of observing event E by counting the total number of outcomes in E and dividing by the total number of possible outcomes in the sample space, S; that is, we say

$$P(E) = \frac{|E|}{|S|}.$$

Since the sample space contains all of the possible outcomes, $P(E)$ is necessarily a value between 0 and 1, inclusive. For instance, the probability that a fair coin lands on heads would be $1/2$ since heads is one of two equally possible outcomes.

Many relative frequency probability problems come down to counting the number of ways an event can occur. It can often be helpful to have a way of organizing the experiment by what outcomes are possible at each step.

Suppose we flip a coin three times and want to determine the probability that exactly two of the flips land on heads. We could start brainstorming a list of what might happen on each of the tosses: HHT, THT, etc. However, if we haphazardly make a list of whatever scenarios pop into our head, we can't always be sure we've

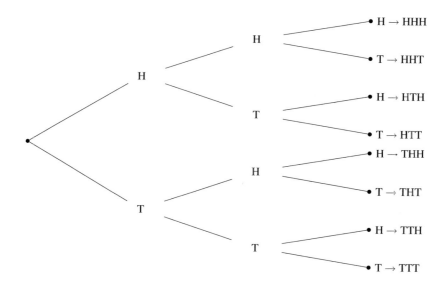

Fig. 2 Tree diagram for tossing a coin three times

thought of all the possibilities. It's nice to have a methodical process; this is where *tree diagrams* can be very helpful.

We start by thinking about the outcomes of the first flip; the coin can come up heads or tails. Let's suppose the coin lands on heads. On the next flip, we will again observe either heads or tails. This gets us to the second branch of our tree—see Fig. 2. Continuing in this way, we find that there are $2 \cdot 2 \cdot 2 = 2^3 = 8$ possible outcomes in our sample space. By following the branches, we can list the outcomes in our sample space as $S = \{HHH, HHT, HTH, HTT, THH, THT, TTH, TTT\}$. If we want event E to be defined as the event that the coin lands on heads exactly twice, we would say

$$E = \{HHT, HTH, THH\}.$$

Since all outcomes in S are equally likely, and event E contains three of the eight outcomes of S, the probability that we observe exactly two heads is

$$P(E) = \frac{|E|}{|S|} = \frac{3}{3}.$$

Exercise 5 Suppose there are three balls—one green, one red, and one blue—in a bag. Use a tree diagram to find all the possible orders of drawing the balls one at a time at random, without replacing the balls between draws. How does adding a yellow ball to the bag change the total number of outcomes? Find an expression for the number of ways to order n differently colored balls.

Exercise 6 Consider an experiment in which you roll two dice.

(a) What is the sample space for the experiment? Hint: list each possible outcome as an ordered pair describing the result of each die. For example, the element (2,4) represents an outcome in which we roll a 2 with the first die and a 4 with the second.
(b) Let E represent the event that you roll two dice with a sum of 9 or greater. List all of the outcomes from the sample space that are also contained in E.
(c) Compute $P(E)$.

All of probability stems from three basic axioms, the first of which we have already discussed:

1. For any event A, $0 \le P(A) \le 1$.
2. $P(S) = 1$, where S denotes the sample space.
3. For any set of disjoint events, $A_1, A_2, \ldots, A_n$,

$$P(A_1 \cup A_2 \cup \cdots \cup A_n) = P(A_1) + P(A_2) + \cdots + P(A_n).$$

The second axiom is asserting that at least one of the possible outcomes in the sample space, S, will occur. The third allows us a way to compute the probability of the union of events that do not share common outcomes. From these three basic axioms, we can derive everything else we need for computing probabilities. For example, suppose we have two complementary events, A and A^c. Based on the rules above, what can you say about $P(A^c)$, assuming you know $P(A)$?

Exercise 7 It is a fact that for any two events A and B,

$$P(A \cup B) = P(A) + P(B) - P(A \cap B).$$

(a) Show that this statement holds by sketching the relevant quantities on a Venn diagram.
(b) How can this statement be simplified if A and B are disjoint events? Hint: consider how Axiom 3 above may be applied here.

2.3   Conditional Probability

Now it's time to introduce the concept of *conditional probability*. We denote the conditional probability of A given B by $P(A|B)$; by this, we mean to calculate how likely event A is to occur *given that B has been observed.*

As an example, let's consider a deck of cards. A standard deck of 52 cards is comprised of four suits—hearts, diamonds, clubs, and spades—each with 13 cards. The thirteen cards of each suit are ranked: Ace, 2, 3, 4, 5, 6, 7, 8, 9, 10, Jack, Queen, and King. We identify each card by its rank and suit. For instance, we could draw an Ace of hearts or a 5 of spades. Note that the heart and diamond suits are colored red, while the club and spade suits are black.

As an example, suppose I select a card at random from a standard deck. If I tell you that the card I've chosen is black, can you tell me how likely it is to be a spade? Put another way, what is the probability that I draw a spade, *given* that I draw a black card? I've already restricted myself to considering black cards only, so I only need to consider what proportion of the black cards are also spades. Put mathematically,

$$P(\text{Spade}|\text{Black}) = \frac{P(\text{Spade} \cap \text{Black})}{P(\text{Black})}.$$

Since $P(\text{Black}) = 26/52$ and $P(\text{Spade} \cap \text{Black}) = 13/52$, we find that

$$P(\text{Spade}|\text{Black}) = \frac{13/52}{26/52} = \frac{1}{2}.$$

In fact, this rule works in general for any two events A and B, assuming $P(B) \neq 0$. We say that

$$P(A|B) = \frac{P(A \cap B)}{P(B)}.$$

That is, in order to find the probability of A occurring once I know B has already occurred, I just need to find what proportion of the set A also resides in B.

We can rearrange this statement to get the following:

$$P(A \cap B) = P(B)P(A|B).$$

Intuitively, this makes sense; in order to observe both A and B, we must first observe B; then, given that we've observed B, we must also observe A. Or, in the language of tree diagrams: first we follow the branch that leads to A, and once we are at A, we follow the sub-branch leading to B. Note that we can also write this in reverse:

$$P(A \cap B) = P(A)P(B|A).$$

This leads us to another interesting question. Are there any scenarios in which $P(B|A) = P(B)$? That is, is it possible that knowing that A has occurred gives us no additional information about the probability of B occurring? Yes! To investigate, we need the concept of independence.

Definition 5 Assume that $P(A)$ and $P(B)$ are both nonzero. We say A and B are *independent* if knowing that one has occurred does not change the probability of the other occurring; that is, if $P(B|A) = P(B)$, or equivalently, if $P(A|B) = P(A)$. If two events are not independent, we say that they are *dependent*.

Exercise 8 We previously determined that $P(A \cap B) = P(A)P(B|A)$. If A and B are independent, how does this impact our rule?

Exercise 9 For each of the following scenarios, describe events A and B using the appropriate terminology: independent/dependent, disjoint, complementary, etc.

(a) You roll two dice. Let A be the event that both are even and B be the event that both are odd.
(b) You roll two dice. Let A be the event that both dice yield the same number and B be the event that the dice differ in value.
(c) You roll a red die and a blue die simultaneously. Let A be the event that the red die is a 3 and B be the event that the blue die is a 4.
(d) You roll two dice. Let A be the event that the sum of the dice is 7 and B be the event that one die is a 6.

We can combine our discussion of conditional probabilities and our previous work with tree diagrams to analyze a number of more complex scenarios. For instance, you'll likely take a number of diagnostic tests for various medical reasons over the course of your life. If a test for some condition comes back positive, you will be most interested in whether or not you actually have the disease, since false positives are possible. Often we know the accuracy of the test in advance; that is, we know how likely the test is to correctly identify the presence of the disease given that we have the disease. In this scenario, our interest lies in finding the conditional probability in the opposite direction; that is, given that we receive a positive test result, what is the probability we actually have the disease?

Let's get at this via a Venn diagram again. If we let A represent the event that we have the disease, and B be the event that we receive a positive test result, we are interested in finding $P(A|B)$, and thus we can compute this by dividing $P(A \cap B)$ by $P(B)$. But how do we find $P(B)$? Note that there are only two options with regards to our having the disease; either we have it or we don't. That is, our sample space can be divided into disjoint sets A and A^c as seen in Fig. 3. The event B, representing a positive test result, can overlap with either space. When B overlaps with A, this represents a correct test result. When B overlaps with A^c, we have a false positive; that is, our test comes back positive even though we don't have the disease. Since all of B is contained in one of these two spaces, we can say that

Fig. 3 Venn diagram
showing division of event B
over the sets A and A^c

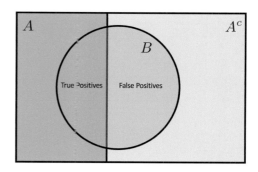

$$P(B) = \underbrace{P(B|A)P(A)}_{\text{True Positives}} + \underbrace{P(B|A^c)P(A^c)}_{\text{False Positives}},$$

since either B occurs given that A is true or B occurs given that A^c is true—there are no other options assuming that B occurs. Substituting this into our conditional probability denominator, we arrive at the following:

$$P(A|B) = \frac{P(A \cap B)}{P(B|A)P(A) + P(B|A^c)P(A^c)}.$$

Let's make this less abstract and consider an example. Note that we are making a variety of simplifying assumptions for the purpose of this example. Suppose 1 in 10,000 people have a rare disease and the diagnostic test for this disease has a 90% accuracy rate. What does this mean? The accuracy rate reports the percentage of time that the test returns the correct result. Thus, if a person has the disease, the test should return a positive result 90% of the time and a negative result 10% of the time. If a person does not have the disease, the test should correctly return a negative result 90% of the time but return a false positive 10% of the time.

We can use a tree diagram—see Fig. 4—to find the probability of actually having the disease *given* a positive test result. Our first branch is based on the prevalence of the disease. The probability that a person has the disease is 1/10,000, or 0.0001, while the probability they do not is 9999/10,000, or 0.9999. Our second set of branches outline our conditional probabilities. The conditional probability of getting a positive test result given that our person actually has the disease is 0.90 while the conditional probability of a negative result is 0.10. Assuming the person does not have the disease, the conditional probability of receiving a positive test result is 0.10, with a negative result probability of 0.90.

Suppose that you take the test and receive a positive result. You are now situated in one of two possible scenarios: either you tested positive and have the disease, which occurs with probability $0.0001 \cdot 0.90 = 0.00009$, or you tested positive but don't have the disease, which occurs with probability $0.9999 \cdot 0.10 = 0.09999$. To find the probability that you have the disease *given* that you tested positive, we can utilize our conditional probability formula derived above:

Fig. 4 Tree diagram for positive test results

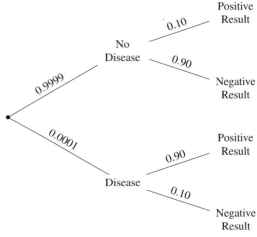

$$P(\text{Disease}|\text{Pos.}) = \frac{P(\text{Disease} \cap \text{Pos.})}{P(\text{Pos.}|\text{Disease})P(\text{Disease}) + P(\text{Pos.}|\text{No Disease})P(\text{No Disease})}$$

$$= \frac{0.90 \cdot 0.0001}{0.90 \cdot 0.0001 + 0.10 \cdot 0.9999}$$

$$= 0.000899.$$

Thus, even if you receive a positive test result, there is only a 0.000899 probability that you actually have the disease—less than a 1% chance! This result is often quite surprising, and may occur when the disease is particularly rare, lending itself to a high number of false positives as compared to true positives. We have to be careful when interpreting reported accuracy. A common misconception is that a 90% accuracy rate means that if you receive a positive test result, there is a 90% chance you have the disease. That conclusion is off by a factor of over 10,000, in this particular case!

When constructing a tree diagram to aid in computing the probability of a sequence of events, note that the first set of branches are based on the random circumstance with an unconditional probability while the second set of branches contains the associated conditional probabilities. We find the probability of the sequence of interest by multiplying the probabilities along the associated branches.

Exercise 10 Suppose your school's bookstore orders textbooks from three different suppliers. 30% of their books come from supplier A, 50% from supplier B, and the final 20% from supplier C. Books from supplier A have a 2% chance of arriving in damaged condition. Books from suppliers B and C each arrive damaged only 1% of the time.

(a) What is the probability that a randomly chosen book arrives in damaged condition?
(b) Given that a book arrives in damaged condition, how likely is it that it came from supplier A?

(Hint: Draw a tree diagram to outline all of the possible outcomes and associated probabilities for each.)

2.4 Random Variables and Probability Functions

We finish our treatment of basic probability with a brief discussion of random variables. We note that the main purpose of this section is to motivate the use of random sampling in Sect. 3.1. As such, we gloss over some details, but point the interested reader to any of [1, 2, 4, 7, 11] for further information.

A *random variable* can be used to represent an as-yet-unknown outcome of an experiment. For instance, when rolling a die, we can let X represent the value of the die, and say that the possible values of X are $\{1, 2, 3, 4, 5, 6\}$. Alternatively, we could let random variable Y represent the height of a randomly chosen fourth-grade student in inches, where we say that Y can take on any real number value in the continuous interval $[46, 60]$. (Note: we are making a large assumption here that no fourth-grade students have heights outside of this interval. Often, such assumptions must be made, and care must be taken to define the possible values of your random variable to accurately represent reality to the greatest extent possible.) We are often interested in determining the probability that a random variable takes on a certain value or resides within a particular interval of values.

There are two types of random variables: those with a countable number of distinct values (i.e., X = number rolled on a die) and those with an uncountable number of possible values (i.e., Y = height of randomly chosen fourth-grader). We call these *discrete* and *continuous* random variables, respectively. Note that discrete random variables may have either finitely or infinitely many possible values, so long as they are countable; that is, easily ordered according to some one-to-one correspondence with the positive integers. Continuous random variables have infinitely many possible values by definition.

2.4.1 Discrete Random Variables

Our die rolling example is an example of a *discrete* random variable since there are a countable number of possible values. If we assume the die is a fair six-sided die, then each of the possible outcomes should be equally likely. Specifically, since there are six equally likely outcomes, each outcome will occur with probability 1/6. We can use a table to summarize our outcomes and their associated probabilities.

Fig. 5 Probability mass
function for a roll of one die

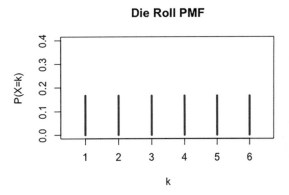

k	1 2 3 4 5 6
$P(X = k)$	$\frac{1}{6}$ $\frac{1}{6}$ $\frac{1}{6}$ $\frac{1}{6}$ $\frac{1}{6}$ $\frac{1}{6}$

Additionally, we can graph a pmf as in Fig. 5, showing a discrete probability for each of the possible outcomes.

The function—or mapping—that assigns probabilities to each possible value of our random variable is called a *probability mass function*, or pmf. The pmf describes how the total probability is allocated among each of the possible values of X. In the case that each of the outcomes are equally likely, as in rolling a single die, the distribution is called a *discrete uniform distribution*. Generalizing our results from Sect. 2.2, we can make the following statements regarding pmfs for discrete random variables:

1. The sum of the probabilities over all possible values of X equals 1; that is,

$$\sum_k P(X = k) = 1,$$

where k ranges over the possible values of X.
2. For each possible outcome k, $0 \leq P(X = k) \leq 1$.

Let's look at an example with a more complex pmf. Suppose you're playing a carnival game where you draw marbles out of a bag. There are two red marbles and two blue marbles in the bag, and you draw three times, replacing your marble after each draw. For each red marble that you draw, you win $1.00. Let X represent the amount of money won.

Exercise 11 For the marble game described above,

(a) What are the possible values of X?
(b) What are the probabilities associated with each of the possible outcomes? (Hint: draw a tree diagram to list all the possible arrangements of the three marbles, and use this to assign probabilities to each potential value of X.)

You might have noticed a similarity between your tree diagram from the previous exercise and our coin flipping example from Sect. 2.2. In fact, they are the same exact tree! Since red and blue marbles are equally likely to be drawn, this scenario is mathematically equivalent to flipping a fair coin. The possible amounts of money you can win are $0, $1, $2, and $3; your pmf should look something like this:

k	0	1	2	3
$P(X = k)$	$\frac{1}{8}$	$\frac{3}{8}$	$\frac{3}{8}$	$\frac{1}{8}$

In graphical form, the pmf for either the marble game or the three-coin-toss example takes the form of Fig. 6.

Let's take this a step further. Though you win $1.00 for every red marble drawn, suppose it costs you $2.00 to play the game in the first place. Is it worth it to play? To answer this question, we need the concept of expected value.

Definition 6 The *expected value* of a discrete random variable is given by the sum of the k possible outcomes multiplied by their associated probabilities:

$$\text{Expected Value} = \sum_k k \cdot P(X = k).$$

This value represents the center of the probability mass function for X.

Fig. 6 Probability mass function for the marble game or a three-coin-toss scenario

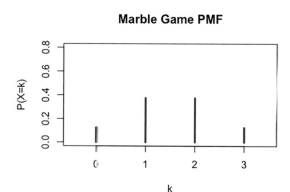

Marble Game PMF

You can think of the expected value as a sort of weighted average. Since we are more likely to observe values of $1 and $2 in our game, we should weight those outcomes more heavily when trying to find the center of our distribution. Let's calculate our expected winnings:

$$\sum_k k \cdot P(X = k) = 0 \cdot P(X = 0) + 1 \cdot P(X = 1) + 2 \cdot P(X = 2) + 3 \cdot P(X = 3)$$

$$= 0 \cdot \frac{1}{8} + 1 \cdot \frac{3}{8} + 2 \cdot \frac{3}{8} + 3 \cdot \frac{1}{8}$$

$$= 1.50.$$

Therefore, if we play the game many times, we expect to win $1.50 on average. Since we have to pay $2.00 to play, this is not worth it to us; we might occasionally get lucky, but on average, we'll lose 50 cents.

Exercise 12 For the marble carnival game, suppose the prize money is $2 for one red marble, $4 for two red marbles, and $8 for three red marbles. How much would the carnival have to charge for game play in order to ensure that they would break even in the long run?

2.4.2   Continuous Random Variables

While a discrete random variable can only take on a countable number of values, *continuous random variables* can take on any of the uncountably many values over some specified interval.

The amount of time it takes to register for classes or the weight of a randomly selected potato at the market are both examples of continuous random variables. Since there are infinitely many values in an interval, it doesn't make sense to talk about the probability that a continuous random variable takes on a specific value. Rather, we think about the probability that X falls within some range of interest. For example, the probability that it takes at least 30 min to register for classes is denoted $P(X \geq 30)$; the probability that a randomly selected potato weighs between 0.5 and 0.8 pounds is denoted $P(0.5 \leq X \leq 0.8)$. These probabilities are computed by measuring the relevant area under the curve of the *probability density function*, the analogue to a pmf for continuous random variables.

The simplest continuous distribution is the *continuous uniform distribution*, the continuous analogue to our discrete uniform distribution. In this case, each interval of the same width has the same probability. Computing the area of interest under the curve is as simple as finding the area of a rectangle since the curve is a horizontal line! The shape of this distribution is completely specified by the endpoints of the interval $[a, b]$; we say that a and b are the *parameters* of the distribution. Recall that for discrete random variables, we required that $\sum P(X = k) = 1$ over all k. The continuous analogue of this rule states that the area under the curve must equal 1. Therefore, for the continuous uniform distribution, the wider the interval chosen,

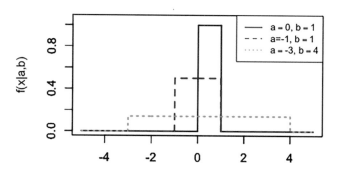

Fig. 7 Examples of continuous uniform distributions. The width and height of the distribution is controlled by the endpoints of the interval—the total area of the rectangle must be 1

the shorter the height of the curve; for an interval of width $b - a$, the height of the resulting pdf must be $\frac{1}{b-a}$. Therefore, the equation for the probability density function of a continuous uniform random variable is given by

$$f(x \mid a, b) = \begin{cases} \frac{1}{b-a}, & a \leq x \leq b \\ 0, & \text{else.} \end{cases}$$

Several continuous uniform curves are shown in Fig. 7 for varying choices of a and b.

Another widely used model for continuous random variables is the normal distribution. The normal curve is bell-shaped and symmetric about the average (or mean) of the population, and is capable of describing a huge number of population characteristics—heights, weights, IQs, etc. The more closely clustered the data is about the mean, the narrower the bell. That is, the shape of the curve is completely determined by the mean, denoted by μ, and the spread—or *standard deviation*, denoted by σ—of the data. The probability density function for the normal distribution with parameters μ and σ is given by

$$f(x \mid \mu, \sigma) = \frac{1}{\sqrt{2\pi\sigma^2}} \exp\left(-\frac{1}{2}\left(\frac{x - \mu}{\sigma}\right)^2\right),$$

where x can take on any value on the interval $(-\infty, \infty)$. This is not a particularly simple function to work with. In fact, integrating this function requires numerical techniques that are beyond the scope of this chapter—see [2,4] for additional details. However, we need not work with this function directly—we'll have R handle that for us in Sect. 3. For now, all we need to know is that changing the value of μ shifts

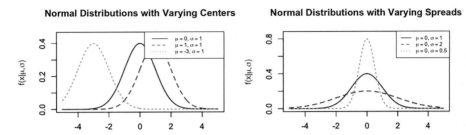

Fig. 8 Examples of normal curves. Changing the value of μ shifts the curve left or right; changing the value of σ alters the spread of the distribution. Note that the area under each curve is 1

the curve to the left or right, while changing the value of σ causes the curve to become either wider and shorter or narrower and taller, preserving the total area of 1. See Fig. 8 for several examples of normal curves with varying means and standard deviations. Note that even though it looks like the curve "zeros out" as you move out toward the extremes in either direction, the horizontal axis actually acts as an asymptote: the curve gets closer and closer but never actually touches the axis. Since the area beneath the curve corresponds to the total probability, it must be the case that the total area under the curve is equal to 1 even though the curve never touches the horizontal axis!

Recall, we said that $P(a \leq X \leq b)$ could be computed by finding the area under the curve between a and b. While this was easy to do for the continuous uniform distribution, the underlying equations for the curves shown in Fig. 8 are actually quite complicated, requiring advanced calculus to compute these areas. However, our next section will focus on exploiting the power of programming software to get everything we'll need from these curves without needing to learn how to do these computations.

Though the continuous uniform and normal distributions have a wide range of applications when it comes to modeling data, we'll use these primarily as a means of simulating "noise"; that is, by overlaying a distribution atop a quantity of interest, we can simulate how that quantity might deviate slightly from its expected value. For instance, if 50 professors are asked to assess a paper and grade it on a 0–100 scale, they might return grades that are centered about 75 but range from 65 to 85 based on personal preference. Section 3.1 will describe how to sample numbers at random from these distributions, and these tools will be used further in Sect. 4 to investigate relevant applications.

3 Introduction to Simulation

In Sect. 2, you learned how to compute probabilities of occurrences of certain events, investigated several popular probability distributions, and worked with more advanced concepts like conditional probability and expected value. In order to incorporate all of those ideas into a computer simulation, we need tools that enable us to do the following:

1. Draw a random number from a given distribution (Sect. 3.1)
2. Check to see if a criterion is satisfied (Sect. 3.2)
3. Continue a task as long as a criterion is met or for a pre-specified number of times (Sect. 3.3)

Though the command names and notation may vary between programs, all simulation software possesses the capability to perform these simple operations. Throughout these sections, our default will be to use the notation and commands of R. Before you proceed any further, you should make sure you have R or another simulation software of choice installed on your computer, and that you have some familiarity with starting new scripts, defining variables, using the calculator functions, and other basic features. For students who choose to use the statistical software, R, we point them to [5, 12] for additional resources for getting started.

Sections 3.1–3.3 will introduce you to a few of the most basic tools of simulation and help you to practice your programming skills. Once you're feeling comfortable using these tools in isolation, you can move to Sect. 3.4 to practice putting it all together.

3.1 Random Number Generation

In general, we want to be able to incorporate an element of "randomness"— or *stochasticity*—into our simulations, to allow for observance of a variety of different outcomes that occur with certain underlying probabilities. In Sect. 2.4, you were introduced to several common probability distributions: discrete uniform, continuous uniform, and normal. How can we randomly generate numbers from each of these distributions for use in simulation?

3.1.1 Sampling from Discrete Distributions

Let's return to our simple coin flip example. We can assign number "1" to "heads," and "0" to tails, and then ask R to select "1" or "0" at random (with equal probability) to simulate a coin flip. The simplest way to do this is using R's `sample` command, which requires that you specify a vector from which it should sample followed by the number of samples you want. For instance, the command `sample(x=c(0,1),size=1)` will return either "0" or "1," with equal probability, representing the outcome of a single coin toss. (Note: the `c()` command you see in the above code means 'concatenate'; it creates a vector out of the given values.) But what if we want to simulate 10 coin tosses? Using `sample(x=c(0,1),size=10)` should do the trick, but we need to make one more adjustment. We'll add the option `replace = TRUE` in order to allow the sample function to choose "0" or "1" more than once in the course of its 10 draws. Here is one possible outcome:

```
sample(x=c(0,1),size=10,replace=TRUE)
```

```
0 1 1 0 0 1 1 0 1 0
```

If we repeat this command a second time, we will almost certainly get a different outcome. If you were to draw a tree diagram outlining all of the possible combinations of heads and tails that could be observed in 10 flips of a coin, you would find that there are 2^{10} different outcomes that we could observe using this single command! This is the beauty of simulation: by incorporating stochasticity into our programs, we can observe a large variety of different outcomes almost instantaneously.

Note: in R, if you want to see the documentation for a particular function—for instance, if you need to look up the order in which the inputs should be specified—you can use the "?" functionality. For example, typing "?sample" into the console and hitting enter will bring up the complete documentation for the sample function.

Exercise 13 Simulate 1000 coin flips. How many heads did you observe? (Hint: assign the result of your toss simulation to a variable called "coinToss," and use the sum(coinToss) command to count the total number of heads.) Repeat your experiment a second time. How many heads did you obtain this time?

Exercise 14 Set up a command to simulate 10 rolls of a six-sided die. How often did you observe each number? (Hint: assign the result of your die rolls to a variable called "dieRolls," and use the table(dieRolls) command to observe how many times each number occurred. Then, use the barplot() command to create a barplot out of your table results.) Repeat your experiment again for 100, 1000, and 10,000 rolls. What do you notice about your bar plot as the number of simulated die rolls increases?

Another useful option in sample is to alter the probability distribution corresponding to your sampling vector (though we stress that this is no longer a *uniform* distribution!). For example, if we wanted to simulate a coin flip for a biased coin that had a 70% chance of coming up heads, we could use the command sample(x=c(0,1),size=1,prob=c(.3,.7)). This command would return "1" roughly 70% of the time, and "0" the other 30%.

Exercise 15 Simulate 10,000 rolls of a biased die, where even numbers are twice as likely to be rolled as odd numbers. Create a bar plot to visualize your results. Does the plot look how you expected?

3.1.2 Sampling from Continuous Distributions

A second distribution that we commonly encounter in simulation exercises is the continuous uniform distribution. Recall, this distribution assumes that all real numbers in the interval $[a, b]$ are equally likely to occur.

In R, we can generate n random numbers from the uniform interval $[a, b]$ using the command runif(n,min,max). For example, to draw four random numbers from the interval [2, 5], we can do the following:

```
runif(n=4,min=2,max=5)
```

```
2.501532 4.323880 2.781780 3.186522
```

Note that each time we run this command, we'll get different results—try it yourself! We may also be interested in where a randomly generated number falls within the probability density function. Let's take our first generated number from above, 2.502, as an example. How likely are we to draw a number less than that from the continuous uniform distribution on [2, 5]? The command punif(q,min,max) can calculate this for us. The "p" stands for "percentile," and measures the percentage of the distribution that falls to the left of the value q. Entering punif(q=2.502,min=2,max=5) returns "0.1671773"; that is, there is a 16.72% chance that we will observe a number less than 2.502 when drawing from the [2, 5] continuous uniform distribution. Figure 9 illustrates this visually using the pdf curve.

We can also go the other direction. Perhaps you're interested in knowing which number represents the 75% mark in your distribution. Use qunif(p,min,max) to find your answer. Here, "q" stands for "quantile"; it will return the number that corresponds to the pth percentile in your uniform distribution. For our uniform distribution on the interval [2, 5], running the command qunif(p=.75,min=2,max=5) tells us that 4.25 is the number such that 75% of our distribution lies to its left, and 25% of the distribution lies to its right—see Fig. 9.

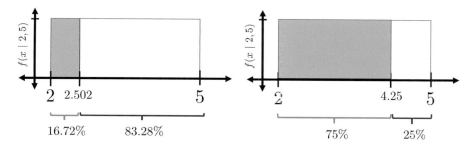

Fig. 9 Finding percentiles and quantiles on the continuous uniform distribution defined on the interval [2, 5]. On the left, we find that 16.72% of the area under the curve lies to the left of 2.502, leaving 83.28% of the distribution to the right of 2.502. On the right, we learn that 4.25 is the number such that 75% of the distribution lies to its left, leaving 25% of the distribution to its right

Just as we looked at `runif`, `punif`, and `qunif` for the continuous uniform distribution, R has built-in functions `rnorm`, `pnorm`, and `qnorm` for the normal distribution. In fact, R has built-in support for many different distributions (both continuous and discrete) under the same naming scheme; for instance, `rexp`, `rgamma`, and `rbinom` are commands for generating random numbers from the exponential, gamma, and binomial distributions, respectively. For students interested in exploring other distributions such as those listed above, we recommend [2, 4]. In this chapter, we will keep our focus on the normal distribution in addition to the discrete and continuous uniform distributions.

In order to use these special distribution commands, we need to specify the parameters of the distribution. For the continuous uniform distribution, this meant specifying a and b, the bounds of your interval. For the normal distribution, we need to specify a mean, μ, and a standard deviation, σ, so that R knows the shape of the probability density function. Thus, the first three inputs in `rnorm` should be the number of samples you want to produce followed by the mean and standard deviation of your distribution, i.e., `rnorm(n,mean,sd)`.

Exercise 16 Generate a random number from the standard normal distribution (the normal distribution with mean 0 and standard deviation 1). Compute the percentage of the distribution that lies to the left of your randomly generated number.

Exercise 17 Find the number that corresponds to the 25th percentile of the normal distribution with mean 10 and standard deviation 3. Using this information—and what you know about the normal distribution—can you find the number that corresponds to the 75th percentile of the distribution without using R?

3.2 "If/Else" Statements

The next tool we need to develop is a way to check whether or not a certain criterion is met or if a particular statement is true. We'll do this using "if" and "else" statements, but first we need to discuss some basic logic checks. Table 1 gives the notation used in R to perform logic checks.

Table 1 Notation for logic checks in R

Notation	Meaning
>	Greater than
>=	Greater than or equal to
<	Less than
<=	Less than or equal to
==	Equal to
!=	Not equal to
&	And
|	Or

As an example, consider the following code:

```
x = 12      #Store the value 12 in "x"
x >= 11     #Make a logic check on the value of x
```

R will return

```
TRUE
```

because the value stored in x satisfies the statement "x is greater than or equal to 11."

The "if" statement asks R to run a logic check. If the logic check returns "TRUE," the next section of code will be run; otherwise, the code contained within the statement will be ignored. The general structure of an if statement is as follows:

```
if( write logic check here ){
    code to be run if logic check is true
}
```

Here's a simple example of the "if" statement in action:

```
if ( x >= 11 ){
    print("x is greater than or equal to 11")
}
```

If I set x = 12 and then run the above code, R will print the statement "x is greater than or equal to 11." If, instead, I set x=5 and then run the above code, nothing will happen, because our logic condition was not satisfied.

Note that you can combine logic checks using & or |. For example:

```
if (x <= 12 & x>=11){
    print("x is a number between 11 and 12, inclusive")
}
```

Note that the logic check in the if statement will only return "TRUE" if both the first part (x<=12) and the second part (x>=11) are true. Similarly, if you use |, the if statement will return "TRUE" if *either* the first part or the second part is true (or both).

What if you want your code to perform one action if the logic check returns "TRUE," but do something else if it returns "FALSE"? We can use an "else" statement to get this behavior. We'll tack an extra request onto our code, sandwiched inside an else statement; this line will only be read if the original logic check is false. For example, if I set x = 5 and then run the code

```
if (x >= 11) {
  print("x is greater than or equal to 11")
} else {
  print("x is less than 11")
}
```

R will print "x is less than 11."

Exercise 18 Write a code snippet that will check if x is equal to 6. If it is, your code should print "This number is 6"; if not, your code should print "This number is not 6."

Exercise 19 Write a code snippet that will (a) generate a random integer x from the set {1, 2, . . . , 10}, and then (b) check if x is an even number. If x is even, divide x by 2. If not, print "x is not divisible by 2." (Hint: In R, x%%y returns the remainder of x divided by y.)

3.3   "While" and "For" Loops

Another useful tool in our simulation toolbox is the ability to repeat a command over and over using a loop. We'll discuss two types of loops here: "while" loops, which cycle for as long as a specified condition is met, and "for" loops, which run for a pre-specified number of iterations.

3.3.1   "While" Loops

The "while" statement is very similar to the "if" statement. Like the "if" statement, it will perform a logic check. However, unlike the "if" statement, it will keep running the code in the brackets until the logic check returns "FALSE." When this occurs, the loop will break and the code will proceed to your next step. The general structure is as follows:

```
while( write logic check here ){
    code to be run as long as logic check remains true
}
```

Here's an example of the "while" statement in action. What is the first negative number in the sequence {100, 99, 97, 94, 90, 85, . . .}? The following code will find it for us:

```
num = 100                      #our starting value
subtract = 1                   #starting subtraction
while(num > 0){
    num = num - subtract       #find the next value
    subtract = subtract + 1    #increment subtraction
}
print(num)
```

In this code snippet, the while loop checks to see whether our number is still positive. If it is, then the code subtracts the next deduction, before updating the value to be subtracted next time. Once the variable num goes negative, the while loop will break and we'll be left with the current value (-5). Try this on your own to confirm, before trying the next example on your own.

Exercise 20 What is the first number in the sequence 1, 3, 9, 27, . . . that exceeds 1,000,000? What is the last number in this sequence that is less than 1,000,000? How many numbers are in the sequence when the while loop breaks?

A final note of caution about using while loops: make sure that your condition will eventually be false! A while loop will continue running until the condition is satisfied, so it's easy to accidentally get stuck in an infinite loop. For that reason, it's often preferable (though not always possible) to use a second type of loop where you specify in advance exactly how many iterations you'd like to perform. This brings us to our next tool.

3.3.2 "For Loops"

"While" loops can be useful tools for repeating a command over and over, but the number of times that command is repeated is contingent upon the logic condition you set in advance. What if you know exactly how many times you want to perform an action? In this case, a "for" loop is more appropriate.

Unlike the "if" and "while" statements, the "for" statement does not perform a logic check. Instead, it repeats the code in the brackets for the number of times specified in the statement. The general structure is as follows:

```
for(statement indicating number of code iterations) {
    code to be repeated
}
```

To specify the number of times for which you'd like to execute a given snippet of code, you can use a statement like i in 1:1000, or some other vector of values. Note: the notation 1:1000 specifies the list of all integers from 1 to 1000. On the first run through the for loop, your program will set i=1. When it repeats for the second time, the value of i will be 2. This will continue until the final iteration, when i=1000. At this point, the program will exit the loop and continue on in your code. Here's an example of the "for" loop in action:

```
x = 10
for(i in 1:5){
    x = x-2
}
```

Note that the value of i does not come into play in this code. Instead, we simply take the value of x and subtract 2 five times, updating the value of x after each reduction. However, we could take advantage of the value of i as well. For instance, if I wanted to be able to print the value of x at each iteration of the previous loop after completing the code, I could do the following:

```
xVals = NA #create an empty variable
x = 10
for(i in 1:5){
    x = x-2
    xVals[i] = x   #save current value of x
```

```
}
print(xVals)
```

Try this on your own. What happens?

Let's try a few more exercises of varying complexity on your own to familiarize yourself with this essential tool.

Exercise 21 Consider the following snippet of code:

```
b = 1
for(i in 1:8){
    b = b + i*b + 1
}
```

(a) What is the final value of b after running the loop above?
(b) Amend the code above to check whether or not b is an even number at each iteration. Create a counter that keeps track of how many times an even number is obtained over the course of the loop.

Exercise 22 Start with the number 1. Use your simulation tools to add 2, then subtract 3, then add 4, then subtract 5, etc. What is the 500th number in this sequence? How many numbers must you add/subtract before you get a number that exceeds 100?

Exercise 23 The Fibonacci series is a sequence of numbers such that each number is the sum of the two numbers that immediately precede it. Typically, the sequence starts with two 1's, and thus takes the form "1, 1, 2, 3, 5, 8,...." Write a piece of code to generate the first 50 Fibonacci numbers. Then, determine which of those numbers are divisible by both 2 and 3.

3.4   Writing More Complex Simulation Code

It's time to put everything together and build more complex simulation code that can confirm the intuition that we develop by working through problems by hand. A helpful way to get started on a complex set of code is through "pseudocode": diagramming the flow of your code without worrying yet about the notation and small details.

Let's use the popular game *Yahtzee* as an example. In *Yahtzee*, a player gets three rolls of five dice per turn. After each of the first two rolls, the player may choose to keep any number of dice and re-roll the rest. In the actual game, there are a number of different outcomes a player hopes to collect over the course of the game (i.e., straight, full house, three-of-a-kind, etc.), but the real prize is a "Yahtzee," or five-

---

Estimating the Probability of Obtaining a Yahtzee via Simulation - Basic Version:

1. For each trial:

    - Roll five dice.
    - Set aside all dice belonging to biggest match set.
    - Re-roll remaining dice.
    - Set aside all dice belonging to biggest match set.
    - Re-roll remaining dice.
    - Check whether or not a Yahtzee was obtained.

2. Count how many Yahtzees were obtained across all trials.

Fig. 10 Basic pseudocode for Yahtzee example

of-a-kind. How likely is it that you roll a Yahtzee over the course of your three rolls? Let's investigate.

Exercise 24 Draw a tree diagram illustrating all the paths to a Yahtzee over the course of your three rolls. (Hint: there are many different ways to define your events of interest—we recommend tracking the "largest number of matches" at any given roll. For example, using this as our metric, your first roll should result in five different branches—"no matches," "at most two matches," "at most three matches," etc.)

Computing the probability of obtaining a Yahtzee over three rolls by hand is actually very difficult and requires a solid foundation in a mathematical sub-field called combinatorics—the interested reader may see [2, 4] for an introduction to combinatorics usage in probability. However, by using simulation, we can get a rough estimate of this probability without knowing the first thing about combinatorics! Let's write out some pseudocode to outline the steps that we'll need to incorporate into our code. We'll show ours in Fig. 10, but you should feel free to write your own, as there are many different ways to implement this task. Let's assume that we want to simulate many trials of three rolls each, so that we can estimate a long-term frequency of obtaining a Yahtzee.

The pseudocode we wrote in Fig. 10 is very bare. We've outlined the basic steps, but have not gone into very much detail about how we'll accomplish them. For some scenarios, this may be more than enough to get you started on your actual code. But if you're tackling a particularly difficult task—or if you're a beginner programmer—it may be helpful to refine your pseudocode a few times to hash out the details and give yourself more guidance when it comes time to actually write the code. For instance, in our first draft we had a sentence that read "Set aside all dice belonging to biggest match set." While that might be enough information if you're explaining game strategy to a new player, it tells us very little about how we'll implement it in the code. In fact, there are several sub-steps hiding within this one sentence! How will we determine which number has the most matches? What if there are two

---

Estimating the Probability of Obtaining a Yahtzee via Simulation - Refined Version:

1. Initialize a variable for keeping track of whether of not we get a Yahtzee each trial.
2. Fix the number of trials we want to run.
3. For each of the simulation trials:

   - Roll five dice by sampling five values from the set $\{1,2,\ldots,6\}$ with replacement.
   - Find the mode (the most frequent value) in the sample. Count how many times it was observed. If two or more modes exist, choose the smallest of the numbers. Reserve all dice belonging to this mode.
   - Re-roll all dice not belonging to the mode by sampling the difference of five and the number in the mode from the set $\{1,2,\ldots,6\}$ with replacement.
   - Recombine the reserved dice and the newly sampled dice by concatenating the two sets. This represents your second roll.
   - Find the mode in the sample. Count how many times it was observed. If two or more modes exist, choose the smallest of the numbers. Reserve all dice belonging to this mode.
   - Re-roll all dice not belonging to the mode by sampling the difference of five and the number in the mode from the set $\{1,2,\ldots,6\}$ with replacement.
   - Recombine the reserved dice and the newly sampled dice by concatenating the two sets. This represents your third roll.
   - Check to see if a Yahtzee was obtained by finding the mode and seeing if it contains five dice. If so, record a "1" in the variable that keeps track of Yahtzees. If not, record a "0".

4. Count how many Yahtzees were obtained in all trials by taking the sum of the vector in which the Yahtzee results were recorded. Estimate the probability of obtaining a Yahtzee by dividing this sum by the total number of trials.

Fig. 11 Refined pseudocode for Yahtzee example

numbers that have an equal number of dice? What does it mean to "set aside the dice" in terms of code? If desired, we could greatly refine our pseudocode—as in Fig. 11—to give ourselves more direction.

Now that we have some idea of what needs to be accomplished and in what order, we can translate our pseudocode into actual code using our step outline to structure our thinking. Our version of the Yahtzee code is shown in Fig. 12. This is by no means the only way to set up this code (or even necessarily the most efficient), but it is relatively simple to understand and follows easily from our earlier outline in Fig. 11. (Note: to use the `mfv()` function in the given code, you'll need to first install the `modeest` package from the "Tools" menu and run the line `library(modeest)` prior to running the code.)

Run this code, or develop your own. Repeat the simulation several times. What do you notice? Do the results agree with your expectation? What happens if you increase or decrease the number of trials? You'll likely discover that there is a dependence of accuracy/consistency upon the sample size used in your simulations. This is a small peek into the field of statistical inference—using information obtained about a *sample statistic* to make inferences about an unknown *population*

```
#YAHTZEE SIMULATION CODE

#Initialize variable to store results
#and number of simulations to run
keepTrack = NA
numSims = 10000

#Start for loop for 10000 simulations
for (i in 1:numSims){

  #ROLL 1: Roll five dice
  roll1 = sample(1:6, 5, replace=TRUE)

  #Find the most frequent value
  #(if tie, choose the first)
  numMost = mfv(roll1)[1]

  #Find how many dice are MFV
  numMatch = sum(roll1=numMost)

  #ROLL 2: Re-roll remaining dice
  firstReroll = sample(1:6, 5-numMatch, replace=TRUE)

  #Combine re-rolled and saved dice
  roll2 = c(firstReroll, rep(numMost,numMatch))

  #Find the most frequent value and number of dice
  numMost = mfv(roll2)[1]
  numMatch = sum(roll2=numMost)

  #ROLL 3: Re-roll remaining dice and combine
  secondReroll = sample(1:6, 5-numMatch, replace=TRUE)
  roll3 = c(secondReroll, rep(numMost,numMatch))

  #Find the most frequent value and number of dice
  numMost = mfv(roll3)[1]
  numMatch = sum(roll3=numMost)

  #TABULATION: Update keepTrack with result
  if (numMatch == 5){
    keepTrack[i] = 1 #we got a yahtzee!
  }
  else {
    keepTrack[i] = 0 #no yahtzee
  }
}
#Print estimated probability of getting a Yahtzee
print(sum(keepTrack)/numSims)
```

Fig. 12 Simulation code for Yahtzee example. Code requires use of modeest package for R

parameter. We point the interested reader to [1, 6, 7, 11] for additional information about basic statistical inference, and to [2, 4] for a more theoretical discussion of uncertainty and the construction of confidence intervals for estimating population parameters.

By now, you've seen that simulation can be a useful tool for confirming your intuition. Additionally, it can be extremely helpful for estimating results in scenarios where we cannot work a problem by hand efficiently.

Let's consider another scenario for practice. In Professor Lewis' math modeling class, a portion of the course grade comes from completing a series of extra activities chosen by the student. These activities include writing reflections on guest lectures (2 points per reflection), completing programming labs (each lab worth 2 points), writing formal solutions to modeling scenarios discussed in class (4 points each), or meeting with assistants in the college writing center to get feedback on project reports (3 points per visit). Students must complete 15 points worth of activities to receive full credit.

Exercise 25 Suppose a student chooses activities to complete at random with no regard to the number of points they are worth. What are the theoretical minimum and maximum number of activities needed to reach the 15-point threshold? What is the *expected* number of activities that the student will have to complete to get full credit? Use your intuition and skill set from Sect. 2 to answer these questions. Then, develop code to simulate this scenario by first writing out a pseudocode, and then adapting it into the programming language of your choice. Run your simulation 1000 times, construct a histogram of the number of activities needed in each trial, and comment on the distribution.

Exercise 26 To encourage students to pursue a wide variety of activities, Prof. Lewis puts a limit of two activity submissions per category. Revise your simulation from the previous exercise to account for this cap. How does this change your distribution?

---

4   Suggested Research Projects

There is really no limit to the number of real-life scenarios that can be modeled via simulation. In what follows, we present three different scenarios of interest that all utilize the tools presented in this chapter, all revolving around the theme of choosing a winner. Each scenario can be adapted to fit the interests of the student investigator. In all cases, the student should use the skills gained in Sect. 2 to develop some intuition about the way in which they expect the scenario to play out; then, code can be developed to investigate and analyze the wide range of outcomes that may be observed.

4.1 Scenario 1: Objective Ranking

We begin by considering scenarios in which winners are chosen objectively; that is, there is a clear winner and a clear loser. Specifically, our following prompts revolve around the FIFA World Cup, but these prompts could easily be adapted to other similar scenarios to suit the student's interests (i.e., NBA March Madness bracket selection, identification of NFL Superbowl playoff teams, etc.).

In the Fédération Internationale de Football Association (FIFA) World Cup Championship, played once every four years, teams representing countries from across the globe face off in an elimination-style tournament to crown a winner. To begin, the 32 qualifying teams are sorted into eight groups of four teams apiece for the group stage. Each group plays a "round-robin' mini-tournament, where each team plays matches against the three other teams for a total of six matches per group. During this stage, a team gains three points for each win, one point for a draw, and zero points for a loss. After all six matches, the top two point-scorers in each group advance to the knockout rounds. In determining the top two teams from each group, the tie-breaker is the total goal differential across all matches played, followed by the total number of goals scored across all matches played, if necessary. Once into the knockout stage, teams are eliminated in one-off matches until only two teams remain for the final game. Draws are not allowed in knockout matches—teams will play into overtime or even penalty kicks to ensure a winner. (Note that the FIFA Women's World Cup typically begins with 24 teams—the top two teams from each of six groups plus the top four third-place teams across all groups advance to the knockout rounds).

Throughout the tournament, predictions abound among fans and in the media. How will a particular fan-favorite team fare? How likely is it that the #1 and #2 ranked teams going into the tournament will actually play each other in the final match? What are the chances that a highly ranked team gets knocked out in the group stage by an underdog? We can use probability and simulation tools to model the likelihood of observing different tournament outcomes.

Challenge Problem 1 Suppose that each team has a 50/50 chance of winning any game in which they play. How likely is it that Team A will make it to the final match? Compute your answer by hand; then, develop simulation code and run 1000 tournament simulations to estimate this probability. Compare your simulated results to the answer you computed by hand.

Challenge Problem 2 Suppose that each team has a 50/50 chance of winning any game in which they play. On average, how many games should Team A expect to play in the tournament? Compute your answer by hand; then, refine your simulation code to estimate the average number of games played over 1000 tournament simulations.

Research Project 1 In actuality, teams do not have a 50/50 chance of winning every game they play. Implement a scheme in which each team enters the tournament with a ranking of 1–32, with higher-ranked teams being more likely to win against lower-ranked teams. How likely is it that the final match is played by the two highest-ranked teams going into the tournament? Investigate how this probability depends on different factors, including the way in which teams are seeded into groups in the group stage and your outlined scheme for determining the probability of each team winning their matches.

4.2   Scenario 2: Subjective Ranking

Sports like soccer, basketball, and football are relatively straightforward—fishy referee calls aside, there is typically a clear winner. But what about more subjectively judged sports like figure skating or gymnastics? Or what about scenarios in which judges are asked to score a musician's performance, or a work of art, or a paper submission? Subjective judging opens up a whole new world of potential issues including the opportunity for variability of scores between judges or judge bias. How can we model these issues in our simulations to allow us to assess the likelihood of certain outcomes?

For example, let's consider a gymnast competing on the balance beam. Suppose her "true" score is a 14.750; this is the objective version, if such a thing could exist. However, if we showed her routine to five different judges, we might receive marks of {14.250, 14.775, 12.500, 14.825, 14.500}. Some of this could be attributed to "noise"; in subjective judging, there is bound to be some disagreement (particularly with regards to an execution score), and so we are not particularly troubled by seeing the scores vary a bit within a reasonable range. But what about the 12.500 from Judge 3? In comparison to the other marks, this seems remarkably low. We might surmise that this judge is biased against this particular gymnast, whether unconsciously (perhaps Judge 3 just prefers a different style), or consciously (perhaps Judge 3 has a nefarious plan to tank this gymnast so that the gymnast representing the judge's home country emerges the victor). We've seen this sort of scandal before; in the 2014 Sochi Winter Olympics, the gold medal in women's figure skating was awarded to Adelina Sotnikova of Russia over the favorite, South Korean skater Yuna Kim. Sotnikova's performance was widely regarded as inferior to Kim's, and there was suspicion that the panel of judges—four of the nine of whom were Eastern European—had thrown the event her way. Due to the anonymous nature of figure skating judging during that Olympics, this could never be confirmed.

We can use the skills gained earlier in this chapter to represent these issues in subjective judging within our simulations. For instance, to allow for a range of scores from judges, we might randomly generate scores from a normal or uniform distribution centered at the "true" score with some specified spread. To incorporate bias, we might shift that distribution to the right or left for a particular judge, or perhaps even use a skewed distribution that makes it more likely that they will select a score from one end of the distribution over the other.

Research Project 2 Develop a simulation code to represent two competing gymnasts (or skaters, musicians, artists, etc.) for a certain event. Assign each competitor a "true" score and select their given scores based on a distribution centered at that true score with a spread which you define. Investigate how likely it is that the gymnast with the lower "true' score is awarded a win over the other, and how this probability depends upon the type of distribution used, the centers of the distributions, and the values assigned for the measure of spread.

Research Project 3 Expand your investigation from the previous research prompt to include bias. If there are multiple judges whose scores are all averaged together, how drastic does the bias on a single judge need to be to overturn the results? Find a way to quantify the dependence of the result on (a) the difference between the "true" scores for the two athletes, (b) the amount of variability in scores among judges, (c) the level of bias present for a single judge, and (d) combinations of (a)–(c) working in tandem.

4.3 Scenario 3: Comparing Voting Methods

So far, we've looked at building simulations to model scenarios with both objective and subjective judging. But we haven't actually questioned the underlying structure of the tournament or judging protocols themselves! Another benefit that simulation-based research can give us is the ability to compare multiple algorithms for the same scenario to decide which is most fair, most stable, or most efficient. Let's turn our attention to an application that is always relevant: voting.

Many elections utilize a winner-takes-all system, also known as a plurality system. The candidate who receives the most votes is declared the winner, regardless of whether they actually obtained a simple majority of 50%. Though this system is easy to understand and relatively resistant to manipulation, some have criticized it for its inability to allow for emergence of minority opinions. But what are the alternatives?

One option is the idea of rank-order voting, where voters rank all candidates in order of preference. This method was recently used in the New York City 2021 Primary Election, as described at [9]. In the event that their first-choice candidate is eliminated from the race, a voter's vote can be transferred to the next candidate in line. This has the added benefit of assuring voters that their votes will count, and mitigates the problem of vote-splitting, where two similar candidates might split the vote of their supporters and allow for a third, less-desirable, candidate to run away with the win.

While there are many benefits to rank-order voting, there are some drawbacks as well. Above all, we desire a system that will consistently reflect the desires of the electorate, choosing the candidate favored by the majority of voters (or where that is not possible, a candidate that obtains a plurality of the vote). With some rank-order voting schemes, it's possible to observe selection of the "least-unfavored" candidate over the choice of a preferred candidate. Thus, careful comparison of the stability, consistency, and unbiasedness of differing voting systems must be conducted prior to implementation. For readers interested in learning more about alternative voting systems, we recommend getting started with basic descriptions at [3, 8]. For a more comprehensive discussion of the mathematics behind various voting schemes, we recommend [10].

Research Project 4 One type of rank-order voting system is instant-runoff voting (IRV). In this scheme, voters rank all candidates in order of preference. If one candidate receives over 50% of the vote, they are declared the winner. Otherwise, the candidate with the fewest votes is eliminated, their votes are redistributed to their supporters' next choices, and the votes are re-tallied. This system of elimination continues until someone obtains a majority of the votes. Suppose you are running an instant-runoff election with N candidates. How many different ballot-ranking orders are possible? Select probabilities for each of the possible ballot rankings, and develop code to simulate the election. How does the outcome depend upon your defined probabilities? (Hint: to make the scenario more realistic, you might align your fake candidates along a liberal-to-conservative spectrum, and assign probabilities according to what preference orders you might expect to observe most often in reality.)

Research Project 5 Conduct a thorough comparison of three different voting schemes: plurality voting, IRV as described in the previous research prompt, and another variation of IRV in which if no one obtains a majority of the vote in the first round, only the top two candidates advance to the runoff. How sensitive are the results to the set of probabilities you defined for each ballot-ranking possibility? Are there certain ballot-ranking probabilities that result in contradictory results between the three schemes even if all are given the same set of submitted ballots?

The research projects we have outlined here represent only the tip of the iceberg when it comes to the types of problems that can be investigated using probability and simulation. We encourage students and mentors to design their own variations on these problems and adapt them to their own interests.

References

1. Brase, C.H., Brase, C.P.: Understanding Basic Statistics. Cengage (2018).
2. Devore, J.L., Berk, K.N.: Modern Mathematical Statistics with Applications. Springer (2012).
3. Types of Voting Systems. FairVote. https://www.fairvote.org/types_of_voting_systems. Cited 13 Dec 2020.
4. Grinstead, C.M., Snell, J. L.: Introduction to Probability. American Mathematical Society (1997).
5. Grolemund, G.: Hands-On Programming with R: Write Your Own Functions and Simulations. O'Reilly Publishing. https://rstudio-education.github.io/hopr/.
6. Heumann, C., Shalabh, M.S.: Introduction to Statistics and Data Analysis with Exercises, Solutions, and Applications in R. Springer (2016).
7. Illowsky, B., Dean, S.: Introductory Statistics. OpenStax (2013). https://openstax.org/details/books/introductory-statistics.
8. Alternative Voting Systems. National Conference of State Legislatures. https://www.ncsl.org/research/elections-and-campaigns/alternative-voting-systems.aspx. Cited 13 Dec 2020.
9. Ranked choice voting. NYC Board of Elections. https://vote.nyc/page/ranked-choice-voting. Cited 5 Aug 2021.
10. Saari, D.: Chaotic Elections! A Mathematician Looks at Voting. American Mathematical Society (2001).
11. Utts, J.M., Heckard, R.F.: Mind on Statistics. Cengage (2015).
12. Zuur, A.F., Ieno, E.N., Meesters, E.H.W.G.: A Beginner's Guide to R. Springer (2009).

Modeling of Biological Systems: From Algebra to Calculus and Computer Simulations

Alexander Dimitrov, Giovanna Guidoboni, William Hall, Raul Invernizzi, Sergey Lapin, Thomas McCutcheon, and Jacob Pennington

Abstract

In this chapter we present a survey of mathematical modeling approaches to biological problems. We approach the modeling efforts through several mathematical techniques: from basic algebra, already studied in middle- and high school, to calculus, at the transition of high school and college. Throughout the chapter, we rely heavily on computer simulation techniques with which we investigate different aspects of these models and their relationships with the studies systems, illustrating the modern approach of computational thinking. We have presented versions of these problems and models in our own courses in mathematics for non-majors, targeted to first- and second-year colleges students who are interested in biology and life sciences. Many students have told us that they found the approach engaging and that it increased their appreciation of mathematics as an applied science. The problems and project that you will see here are selected from areas of population biology (populations of squirrels, insects, flies, and wasps); use of medical sensing and modeling to detect the blood pressure and heartbeat in patients; and epidemiology, modeling the spread

A. Dimitrov (✉) · W. Hall · S. Lapin · J. Pennington
Washington State University, Pullman, WA, USA
e-mail: alex.dimitrov@wsu.edu; w.hall@wsu.edu; slapin@wsu.edu; jacob.pennington@wsu.edu

G. Guidoboni
University of Missouri, Columbia, MO, USA
e-mail: guidobonig@missouri.edu

R. Invernizzi
Politecnico di Milano, Milan, Italy

T. McCutcheon
University of Sussex, Brighton, UK

E. E. Goldwyn et al. (eds.), *Mathematics Research for the Beginning Student, Volume 1*, Foundations for Undergraduate Research in Mathematics,
https://doi.org/10.1007/978-3-031-08560-4_7

of infectious diseases, from Ebola in West Africa, with an option to study the COVID pandemic in 2020–2021.

Suggested Prerequisites *The sections are ordered in increasing difficulty of conceptual ideas and technical tools. You do not need to read or try them in the indicated order, but if we use concepts from prior section, look them up as needed. In the first two sections we expect acquaintance with the ideas of basic algebra, in particular, with the more advanced concepts of sequences and relationships between their terms. The subsequent two sections start using concepts from differential calculus. Some familiarity with that field would be helpful, although we do provide some guidance. All problems use some form of computational thinking, which involves computer programming. Initially that is in Excel, assuming familiarity with functions and formulas. Then we transition to modeling tools that are used more broadly by practitioners, like Octave and Matlab. We provide links to tutorials for the use of these systems.*

1 Introduction

Mathematical models allow us to pause and reflect on all the "what ifs" and "I wonders" we might have about our universe. In this chapter, you will read about several different ways mathematical models are used to answer questions. Questions like, how to measure someone's heart rate without touching them, how two insect populations relate to crop yield, and more!

1.1 A Description of Mathematical Modeling

Mathematical modeling is a cyclic and iterative process—a delicate dance between making and challenging assumptions. When building models, there is rarely a "right" answer—only answers that stand on specific assumptions are supported by mathematical logic and principles and help us generate productive knowledge in the given context and setting. To help make sense of what can be a difficult process to explain and understand, we frame the modeling process in four steps.

Step 1: Goals, Questions, and Assumptions
What do you want to know? What do you already know? Predictions and analyses stemming from models are only as valuable as the accuracy of the assumptions. We must think carefully about any assumptions we make and the criteria outlined in problems we are given to solve.

Step 2: Build a Model
How can you utilize mathematics (and other fields) to move from your assumptions to the answers to your questions? What areas of mathematics will be most useful? From what subjects can you draw inspiration?

Step 3: Apply your Model
Use the model to answer questions and meet initial goals.

Step 4: Assess and Revise your Model
Are the results/implications reasonable? What happens when you alter various assumptions? Reflect on whether the assumptions posed are accurately incorporated and whether there might be additional constraints to be considered. Can the model be more accurate or efficient? Does it need to be?

We do not mean to imply that you should always proceed linearly from Step 1 through to Step 4. Sudden realizations about the mathematics involved or the phenomenon to be modeled may send you back to a specific part of your model immediately. New data from the field may stop you in your tracks to revise essential assumptions. However, the general structure and spirit of these steps supports an iterative approach to mathematical modeling in which answers are not "right," but well-supported. These steps and our general approaches are not unique to these projects. They parallel directly George Polya's four principles of problem-solving [1] and are found embedded in the description of mathematical modeling from the Society for Industrial and Applied Mathematics [2]. Look for this cycle within your own pursuits!

1.2   Building a Model with Bias

There is inherent bias in mathematical modeling. Our decisions regarding what to study and how to go about studying it are subject to our bias as people. Bias is not something that can be temporarily suspended or minimized—it is inherent to us. Acknowledging our bias and exploring the implicit bias in data does not diminish our results, it strengthens them by allowing us to situate our results appropriately with full awareness of the shortcomings and advantages of the model. As an example, medical research has historically been done exclusively on adult male subjects. In 2015 the United States Food and Drug Administration mandated that clinical trials include female participants. The impact of that methodological decision includes we are still today learning how differently women experience medical emergencies such as heart attacks [3]. This bias was inherent within society as well as within the research community and thus not accounted for in making recommendations.

1.3   A Note About Computer-Based Simulations

All files with instructions for computer simulations are available for download at the GitHub repository typically used for such tasks [4]. We typically develop and share our code in this manner, although the use of repositories like GitHub becomes more obvious after much more experience with coding. For now, just take our word that it is a good idea to learn how to use them.

The first two sections illustrate the use of computer-based simulations for mathematical modeling using simpler and more widely accessible platforms like Microsoft Excel and Google Sheets. Those platforms have the benefit of familiarity and ease of entry, but they are severely limited when problems scale up in size and complexity, becoming progressively more difficult to use, if at all.

Not surprisingly, the professional math modeling community is typically using different platforms in our day-to-day work. They tend to be more difficult to learn and master but allow us to tackle much more complex problems. Sections 4 and 5 present examples in two of these platforms: the commercial MATLAB [5] and the open-source GNU Octave [6]. They have similar capabilities mainly because Octave was specifically created to be compatible with MATLAB and use MATLAB commands and structures without many changes. The main difference currently is that MATLAB provides a nice integrated environment in which one can mix text, computations, and visualizations, while Octave still uses text-based work environment in which results have to be collated to separate documents with some additional effort. One emerging platform which bridges some of the gap between integrated ease of use and open source/free accessibility is Python in Jupyter Notebook [7, 8]. We note that in passing but will not provide additional details in this chapter.

Since both MATLAB and Octave are based on a text-based programming model, we usually direct our students to online tutorials which provide them the foundations of working with these systems and provide them with additional online materials to expand their expertise should they want to. We believe that the basic tutorials provide sufficient material so students can start exercises and later extend their knowledge in the context of problems on which they are working and engaged. MathWorks, the company behind MATLAB, provides a nice free introductory tutorial [9], which awards certificates to students after they finish it. We usually assign that tutorial as homework on our first day of class and collect the certificates. The same tutorial will provide basic familiarity with Octave as well, but there will be differences in the user interface. As an open-source product, Octave does not have the polished tutorials of MATLAB, but there is still a lot of documentation in the form of a Wiki, and several good quality video tutorials, comparable to MATLAB's Onramp [10], that can serve the same purpose.

We recommend that, before you attempt Sects. 4 and 5, you install MATLAB or Octave on your computer, engage with the tutorials, while trying to replicate what they suggest on your system, and continue with the those sections after you finish those tutorials.

1.4   Active Learning

Learning is not a passive activity. To sit and read the words that follow will impart knowledge, no doubt. But to learn about mathematical modeling, one must actively model. Research in the teaching and learning of mathematics, science, and engineering shows consistently that active learning in the classroom leads to improved student learning [11]. When reading a text like this, you are your own teacher in many ways. Teach yourself well! We encourage you to try out the examples provided and explore in new directions. Alter parameters and think about the implications. Adjust assumptions to fit a new space and play with the ideas as you read. This is active learning (and reading!). This philosophy also defines in part the structure of this chapter and its section. Unlike the majority of chapters in this book, we chose to not initially develop "all necessary math" and then present you with challenges and research problems in which to use "it." We chose to state a potential problem relatively early in the corresponding section, then discuss some possible mathematical approaches to it and provide references for specific tools that the reader can follow as needed. This approach allows us to follow our favorite learning strategy, which we use in our work as modelers as well: a combination of foundation learning of general mathematical structures and "just in time" learning of specific mathematical tools that could be of use to a particular problem.

2   Grey Squirrels in Six Fronts Park: Modeling a Changing Population

Section Author   William Hall

Suggested Prerequisites   *Algebra, sequences, difference equations*

In this section, we build and use a mathematical model using spreadsheet software to explore how a population of grey squirrels changes over time given various parameters and observations. This problem is fictional—it was built for the purposes of instruction so that the mechanics of translating the problem into the mathematical model can be centered in the discussion. In the projects that follow, the contexts and mathematical models are authentic and the principles are illustrated outside a fictional park.

2.1   The Problem

Problem 1   A population of grey squirrels is being tracked within Six Fronts Park. In a recent survey of park wildlife, it was estimated there were 1000 squirrels in the park. The squirrel population has been increasing as of late and city officials have asked for a report. Specifically, they want to know when they can expect the squirrel population to reach 2000.

2.1.1   Step 1: Goals, Questions, and Assumptions

We are asked to estimate when the population of grey squirrels will reach 2000. We are told to assume the squirrel population is increasing by 20% each year.

2.1.2   Step 2: Build a Model

Many different models can be built to approximate the given situation and the reader is encouraged to explore those that come to mind. Here, we will use Microsoft Excel to provide a numerical approximation, using difference equations to solve what would be called an initial-value problem in differential equations; however, you will not need experience in differential equations for this activity. See Sects. 4 and 5 for more details on difference equations and differential equations. First, we need to translate our assumptions into mathematical equations and expressions that can be used in our model. We will use $S(t)$ to represent the population of grey squirrels t years from today. We know that we can assume there to be 1000 squirrels in Six Fronts Park today. This is represented in our model as $S(0) = 1000$, where $t = 0$ represents the present day. Assuming the growth rate of the squirrel population is proportional to the squirrel population itself and not explicitly dependent on time, we use $R(S)$ to represent the growth rate of the population of grey squirrels for some given population. Note that while S is a function of t, R is a function a S. Assuming a growth rate of 20% per year in the context of the squirrel population can be understood as: if there are 1000 squirrels in the park, the population will be growing at a rate of $1000 \cdot 0.2 = 200$ squirrels per year. In our model, we can write this more generally as $R(S) = 0.2 \cdot S$. Given this information, we translate the officials' question ("When will there be 2000 squirrels?") to our mathematical notation: Find t such that $S(t) = 2000$.

What makes this problem challenging (and interesting too!) is that as soon as more squirrels are added to the park, the growth rate increases. In our model: as S increases, so too does $R(S)$. For example, if there were 1200 squirrels in the park, the growth rate would be $1200 \cdot 0.2 = 240$ squirrels per year. How do we approximate a population in flux with a growth rate that changes too? One answer: approximation. If we assume that the growth rate remains constant for some finite interval (e.g., one month, three months, six months) we can approximate how many new grey squirrels the model predicts there will be within that time interval. We can do this several times, updating the rate of change as we approximate the accumulative net change in the number of grey squirrels. A combination of the size of those time intervals and how we approximate the rate of change within each interval will determine how accurate our overall approximation will be. In the next section we will use Excel to apply the model and answer the officials' question.

2.1.3   Step 3. Apply the Model

The method showcased in this section relies on assuming non-linear growth can be approximated using linear functions over finite time intervals. To find an exact solution, we would need to consider infinitely small chunks of time (and possibly infinitely small squirrels). However, we can approximate the solution with as much accuracy as we like using just spreadsheet software and a few formulas. First, we know that at $t = 0$, the growth rate is 200 squirrels per year, as we calculated earlier. If we assume this growth rate will remain constant over three months, we can approximate that there will be (200 squirrels per year) · (0.25 years) = 50 additional squirrels in Six Fronts Park over three months. Now there are approximately 1050 squirrels in the park. At this point, we re-calculate the growth rate since there are more squirrels and we know that more squirrels means faster growth. Now, the population is growing at 1050 · 0.2 = 240 squirrels per year. After three additional months, (240 squirrels per year) · (0.25 years) = 60 additional squirrels added during that three-month time interval. Thus, six months from now, we can estimate there to be 1000 + 50 + 60 = 1110 grey squirrels in Six Fronts Park.

We can speed up this process with the help of spreadsheet software (many other software options exist for this type of modeling, but we use Microsoft Excel here). We will track time in the first column, the squirrel population in the second column, and the growth rate with the third (using t, S, and R, respectively, as column headings). The first entry will be the given information, that at $t = 0$, the population S is 1000. We also need to make note of the time span over which we have decided to assume a constant growth rate, which in our case is three months (0.25 years). Once you have listed this value, you can use the "define name..." feature to assign a character string you can use to call up this value in formulas later on. In this example, I used "dt" as the name for the cell containing 0.25, where dt represents a small change in time. The benefit here, instead of typing 0.25 repeatedly, is that we can change the value in the cell and any formulas in which we have used "dt" will update to the new value. This enables us to explore how our assumption (the growth rate was constant for some finite time interval) impacts the overall approximation for the population of grey squirrels.

To calculate the growth rate at this point, we utilize a formula in Excel and let the software run calculations. In the R column, in the cell adjacent to the current population, we want to compute 0.2 multiplied by the current population. To use a formula, type the equal sign followed by "0.2*" then you can either click the cell with the current population or type the cell location directly. This is illustrated in Fig. 1.

Moving to the next row, we first use how much time has elapsed (0.25 years) to give the next value for t. To do so, we use a formula in the next cell in the time column to add the value of "dt" we set earlier (0.25) to our previous time ($t = 0$). Next, we update the number of squirrels in the park after this period (0.25 years) by approximating the number of *new* squirrels by multiplying the current growth rate by the value saved as "dt" and adding it to the previous population (see Fig. 2 for formulas).

Finally, to find the new growth rate after 0.25 years have passed and there are now approximately 1050 squirrels in the park, we use the fill-down feature (click and drag the square in the bottom right corner of the cell with the formula you want to apply). With the first two rows now filled in, if you simultaneously select all three cells in the second row of values, you can use the same fill-down feature to generate approximations as far into the future as you desire. See Fig. 3 for Excel formulas.

According to the current model (see Fig. 4), the squirrel population should reach 2000 between 3.5 and 3.75 years from now.

	A	B	C	D
1	t (years)	S (# of squirrels)	R (Squirrels per year)	Time Span (years)
2	0	1000	=0.2*B2	0.25

Fig. 1 Using excel to calculate the growth rate

	A	B	C	D
1	t (years)	S (# of squirrels)	R (Squirrels per year)	Time Span (years)
2	0	1000	=0.2*B2	0.25
3	=A2+dt	=B2+C2*dt	=0.2*B3	

Fig. 2 Formulas for calculating time and population

	A	B	C	D
1	t (years)	S (# of squirrels)	R (Squirrels per year)	Time Span (years)
2	0	1000	=0.2*B2	0.25
3	=A2+dt	=B2+C2*dt	=0.2*B3	
4	=A3+dt	=B3+C3*dt	=0.2*B4	
5	=A4+dt	=B4+C4*dt	=0.2*B5	
6	=A5+dt	=B5+C5*dt	=0.2*B6	
7	=A6+dt	=B6+C6*dt	=0.2*B7	
8	=A7+dt	=B7+C7*dt	=0.2*B8	
9	=A8+dt	=B8+C8*dt	=0.2*B9	
10	=A9+dt	=B9+C9*dt	=0.2*B10	

Fig. 3 Formulas for model of population of grey squirrels

Fig. 4 Model of population of grey squirrels with constant growth rate for 0.25 year periods

	A	B	C	D
1	t (years)	S (# of squirrels)	R (Squirrels per year)	Time Span (years)
2	0	1000.000	200.000	0.25
3	0.25	1050.000	210.000	
4	0.5	1102.500	220.500	
5	0.75	1157.625	231.525	
6	1	1215.506	243.101	
7	1.25	1276.282	255.256	
8	1.5	1340.096	268.019	
9	1.75	1407.100	281.420	
10	2	1477.455	295.491	
11	2.25	1551.328	310.266	
12	2.5	1628.895	325.779	
13	2.75	1710.339	342.068	
14	3	1795.856	359.171	
15	3.25	1885.649	377.130	
16	3.5	1979.932	395.986	
17	3.75	2078.928	415.786	
18	4	2182.875	436.575	

2.1.4   Step 4. Assess and Revise Your Model

What would happen if we assumed the growth rate was constant for one month instead of three? Here we replace 0.25 with 0.083 (approximately $\frac{1}{12}$). Make a mental conjecture about how the solution might change before reading further or changing your own values. Since we assigned "dt" to represent whatever value is in the cell marking our time span, we can change that value and the entire model will update as well. However, you may need to highlight the bottom row and use the fill-down feature to include more rows in the model. In this model, our approximation for when the population will hit 2000 falls between 3.486 and 3.569 years from today. This interval is earlier than the previous one we found. Does that match your conjecture?

2.2   Conclusion and Exercises

In this section, we modeled the population of grey squirrels in a fictional park using numerical approximation methods. There are many ways to alter this model to pose and answer new questions. Below, you will find several exercises the reader can undertake within the fictional space of the park. Elevating to the level of project could include utilizing the modeling techniques here in analyzing publicly available data (e.g., COVID-19 infection/vaccination rates, unemployment rates) to pose and reflect on questions about critical social issues in one's community.

Exercise 1 Imagine you found out there was an error in the initial report and there were only 800 squirrels in the park at $t = 0$. How does this alter the results of your model under the original assumption? When would the population hit 2000 in that case?

Exercise 2 What if a fire caused a significant drop in the population growth rate during the first six months of the second year? How could you use the spreadsheet we built here to model this case?

Exercise 3 Pose your own "what-if" question and then reflect on the results. What do you learn about both the context and your model?

3   Non-contact Cardiovascular Measurements

Section Authors Giovanna Guidoboni and Raul Ivernizzi

Suggested Prerequisites *Algebra, spreadsheet*

3.1   Context

At every heartbeat, blood is ejected from the heart into the cardiovascular system, generating a thrust that is capable of moving our whole body [12]. This motion can be captured via external sensors, such as accelerometers placed in an armchair or load cells placed under the bed posts. Precisely, these sensors capture the motion of the center of mass (which is the unique point of a distribution of mass in space to which a force may be applied to cause a linear acceleration without an angular acceleration) of the human body resulting from the function of the cardiovascular system, thereby enabling health monitoring without requiring direct body contact. This capability is especially useful for elderly individuals and patients suffering of chronic diseases, such as those suffering of heart failure. What does this have to do with math? Can we reconstruct the position of a known mass placed on the bed by elaborating the signals from the load cells? This is what we will explore in this project. We will see that unexpected results may occur. . . and yet Euler in his master thesis in 1771 already explained all about it.

3.2   The Challenge

Research Project 1 Imagine that a doctor one day asks you to help him solve a problem. He explains to you that it would be his intention to monitor the health of the cardiovascular system of a patient that is bedridden for good part of the day. The cardiovascular system, he continues, is made of many parts all working together to make sure that every inch of our body gets the blood and the oxygen that it needs to function (see Fig. 5 for a schematic representation of the main parts of the cardiovascular system). The heart is the main pump, beating about 100,000 times and pumping about 7200 liters (1900 gallons) of blood each day through an amazing network of blood vessels of different shape and size.

As a doctor monitoring patients with cardiovascular problems, it is very important to know whether the patient has a problem in the heart or the blood vessels, because depending on the situation different medicines can be administered or interventions can be pursued.

The doctor says there are already several ways to understand if a person's heart is working well. For example, you can use the electrocardiogram (often abbreviated to ECG), which consists of measuring the electrical activity of the heart using electrodes placed on the skin (usually on the chest of patients) and provides a graph of voltage versus time. Each peak in the graph corresponds to a heartbeat, and other useful information can be gleaned from the minor peaks (see Fig. 6 for a schematic of the peaks in an ECG graph). In addition to the ECG, the *pulse oximeter* is another sensor that is often used for patient

(continued)

monitoring (see Fig. 7). This sensor is placed on a finger, as if it were a clothespin, and is able to measure the heart rate and the amount of oxygen in the blood through the emission and subsequent measurement of light beam from a point.

Our doctor's idea, however, is to obtain data on the whole cardiovascular system simultaneously, namely including both the heart and the blood vessels, and to do so without touching the body of the patient! In fact, both the ECG and the pulse oximeter require body contact and they have uncomfortable wires that reach a computer capable of processing their signals.

So how can we measure the heartbeat of a subject or other information about his/her cardiovascular system, without applying sensors on him/her? It seems impossible! One could take advantage of the fact that the patient is lying on a bed. So, thinking about it, he/she is touching the bed. Is it possible to use this contact to get some useful information by putting sensors on the bed frame instead of on the patient directly? This is what we will experiment together.

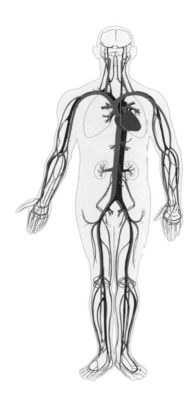

Fig. 5 The cardiovascular system

Fig. 6 Standard
electrocardiogram (ECG)
signal

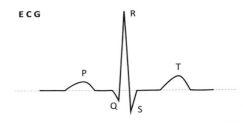

Fig. 7 Pulse oximeter finger
sensor

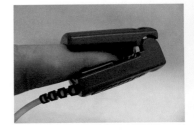

We develop this challenge project in more detail to demonstrate how we work
with broadly defined projects like the ones in this book, how we approach them,
constrain them and define our own smaller project. Subsequent project in this
chapter is meant to be more independent, while still offering some guidance.

3.3   The Initial Experiment

Let us do the following experiment. Take a glass, possibly transparent and with a
wide base. Fill it halfway with water and walk to a place where you can safely lie
down supine (on the back or with the face upward), such as a sofa or a bed. Lie
down carefully, face up, with your head resting on a pillow so that is possible to see
your feet without straining abs. Now, place a glass with water on your chest in such
a way that is stable enough for it not to tip over. Now if you hold your breath for
few seconds and try to remain still you will notice interesting phenomena: the water
moves even if you stay still. One can look closely and see that the surface of the
water moves periodically, perhaps just a little, but it does. But why does this happen
even if we were holding our breath and trying to be immobile?

The answer is: the water move because it is impossible to stay perfectly still,
because we have a beating heart inside us!

Every time that the heart ejects blood into our vessels, it generates a thrust that
moves our whole body in the opposite direction, thereby displacing the water in
the glass. This is Newton's third law, also known as the action-reaction principle:
*to every action there is a reaction that is equal in magnitude and opposite in
direction.* This fact has actually been known for quite some time. Biathlonists or
military snipers, for example, train a lot to pull the trigger taking into account this
phenomenon. In fact, even if they hold their breath, they cannot stop their heart from
making them jump at each beat, so they must learn to synchronize the shot with the
beat.

Fig. 8 Picture of an old prototype of suspended bed

Since the mid-1900s, several scientists were developing instruments to accurately measure these body movements. One of the most popular methods was the "swing bed," which consisted of a rigid board, suspended by means of cables that were attached to the ceiling or to a stiff metal frame (see Fig. 8). They saw that the power of the cardiac contraction that pushed the blood in one direction was able to move the entire bed with the person on it. These movements were, therefore, simple to be measured. In fact, it was enough to fix a felt-tip pen facing down to the oscillating table, and if you make it so that it touches a sheet of paper placed under it, it will then trace the movement of the bed. The signal generated by the body motion in response to the flow of blood in the cardiovascular system is known as *ballistocardiogram*, typically abbreviated to BCG. Since then, different and more precise methods have been developed to measure the BCG, for example, using contactless distance meters such as those laser meters that architects use to measure the size of the rooms of a building, or else by using accelerometers. A system made of a bed suspended with cables is quite impractical to be used in a hospital or at home. So, it is natural to wonder if there is a way to measure the forces that create the BCG using a normal bed resting on the ground rather than a suspended one.

3.4 Weigh a Bed?

Let us imagine that we have 4 equal scales. In practice, in scientific laboratories, bathroom-like scales have more sophisticated substitutes, which are known as "load cells" (see Fig. 9). They may look different than the standard scales, but essentially

Fig. 9 Picture of a load cell
placed under the leg of a bed

they do the same thing: they measure the mass of an object. In addition, these instruments can also be connected to a computer to read the data of the multiple load cells simultaneously and even continuously, obtaining a graph as a function of time. This aspect will come in handy.

If we place one load cell under each bed post, for a total of four load cells, we will be able to take into account the weight that each post unloads on the ground. But what do we do with this apparently useless data?

Let us remember that we are interested in having a measure of how the heart makes the body vibrate in order to measure its beats and possibly obtain other useful information. Is it possible, having only the measurement of the "constraint reactions" of the bed to the floor (i.e., the forces that it transmits to the floor and which, in turn, the floor transmits to the bed by reaction), to obtain the forces transmitted by the person's body to the bed itself? In short, we would like to understand if we can use the load cells to quantify those very same forces that make the water move inside the glass on your chest or make the bed swing when suspended with cables.

The answer is that, thanks to mathematics, it is possible. And it is not very complicated either. Let us see how to get this information:

We draw a diagram of the bed, with the posts at the four corners. We label each corner with a number from one to four. We set a Cartesian reference frame by drawing the x and y along the direction of the bed width and length, as shown in Fig. 10.

We indicate with R_1, R_2, R_3, R_4, the constraint reactions that the bed exchanges with the floor. Basically these are the forces that we read through the load cells. (Note that the direction towards which we draw the forces is not important at the moment. In fact, we can fix the arrows in any direction and, at the end of the computations, the $+$ or $-$ sign that we will obtain will automatically change the directions, if necessary). To simplify the computations, let us consider only one direction at a time, either x or y. We are basically choosing to look at the bed either along its width or along its length and we add together the reactions that act on the same end of the segment, as schematized in Fig. 11.

Fig. 10 Three-dimensional schematic representation of a bed

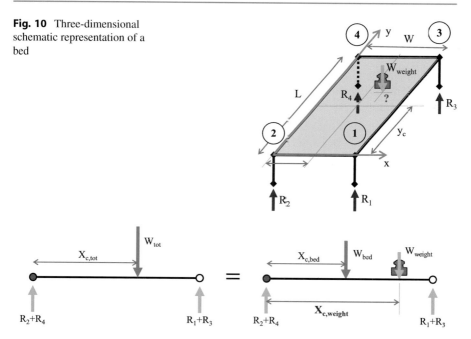

Fig. 11 Cross sectional schematic representation of the forces to which the bed is subjected

Now imagine placing a weight of known mass, say 1 kg, on the bed. We indicate with W_{bed} the mass of the bed, with W_{weight} the mass of the small weight and with W_{tot} the total mass of the bed plus the mass of the small weight. Thus we can write $W_{tot} = W_{bed} + W_{weight}$. We can represent this situation by drawing a downward arrow representing W_{tot}, as schematized in Fig. 11 left, or if we consider the two components related to the weight of the bed and that of the weight separately, we obtain Fig. 11 right (which is anyway equivalent to Fig. 11 left).

The question we ask ourselves now is the following: "Using only the data obtained from the load cells under the bed posts, would we be able to reconstruct the position of the small weight on the bed?" This question will be the important for the next reasoning.

If we are able to reconstruct the position of a small weight on the bed, having only the data from the load cells, it would mean that we are able to measure the deviation of the center of mass of the bed if there is a weight on it. This would lead us to say that, if a person were lying on the bed, we would be able to measure the displacements of the person's center of mass, and these would be caused precisely by the movement of blood inside the arteries. We would have measured the BCG, which provides information about the performance of the whole cardiovascular system!

Returning to our problem, to reconstruct the position of the weight, we use what is called "rotational equilibrium." In physics and mechanics, the moment of a force is the rotational equivalent of a linear force, and to calculate it we just need to multiply the modulus of the force (i.e., its intensity, in N = "Newton" as unit of measure, for example) by the arm (i.e., the distance perpendicular to the force, expressed in m = "meters," for example) from a fixed point that we can choose at will and we will call the fulcrum. As a result, we will get the moment of that force relative to the chosen fulcrum, and it will have Nm = "Newton per meter" as unit of measure.

By enforcing the rotational equilibrium with respect to the left end of the bed indicated with a gray point in Fig. 11, we obtain:

$$[(R_2 + R_4) \cdot 0] - [W_{tot} \cdot x_{c,tot}] + [(R_1 + R_3) \cdot L_{bed}] = 0 \tag{1}$$

$$[(R_2 + R_4) \cdot 0] - [W_{bed} \cdot x_{c,bed}] - [W_{weight} \cdot x_{c,weight}] + [(R_1 + R_3) \cdot L_{bed}] = 0 \tag{2}$$

From the first equation, we can retrieve the location $x_{c,tot}$ of the center of mass along the bed width when the weight on it.

For the second equation we need W_{bed}, $x_{c,bed}$, and $x_{c,weight}$.

Adding the load cell signal and averaging it over time, we can obtain W_{tot}. Because W_{weight} is known, we can obtain W_{bed} by subtraction:

$$W_{bed} = W_{tot} - W_{weight} \tag{3}$$

To obtain $x_{c,bed}$, we need to write the rotational equilibrium for the bed without the weight on it. Let us denote by $R_{1,bed}$, $R_{2,bed}$, $R_{3,bed}$, and $R_{4,bed}$ the load cell reactions that we measure in this case. Thus, we can write

$$x_{c,bed} = \frac{R_{1,bed} + R_{2,bed}}{W_{bed}} \tag{4}$$

Finally, we can combine the two equations to obtain an expression for the coordinate $x_{c,weight}$ identifying the position of the weight along the bed width:

$$\boxed{x_{c,weight} = \frac{W_{tot} \cdot x_{c,tot} - W_{bed} \cdot x_{c,bed}}{W_{weight}}} \tag{5}$$

Repeating the same steps along the y axis, we obtain an expression for the coordinate $y_{c,weight}$ identifying the position of the weight along the bed length:

$$\boxed{y_{c,weight} = \frac{W_{tot} \cdot y_{c,tot} - W_{bed} \cdot y_{c,bed}}{W_{weight}}} \tag{6}$$

CASE	#0	#1	#2	#3	#4
Weight mass [kg]	0.000	1.500	2.000	1.500	1.500
Load cell 1 (R1) [kg]	11.000	11.375	12.500	10.375	11.725
Load cell 2 (R2) [kg]	11.000	11.375	11.500	11.625	10.775
Load cell 3 (R3) [kg]	9.000	9.375	9.000	9.625	9.275
Load cell 4 (R4) [kg]	9.000	9.375	9.000	9.875	9.725

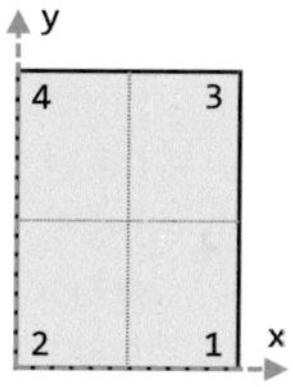

Fig. 12 Table reporting the mass of the weight positioned on the bed, and the magnitude of the constraint reactions of the bed posts measured by the load cells. Note that there are 5 different cases associated with 5 different weights positions. Case #0 is the reference case, where there is no extra weight on the bed

3.5   Let Us Get Our Hands Dirty!

Exercise 4 Let us now have some fun and use what we have learned so far to reconstruct the position of a small weight placed on a bed, given the readings from four load cells placed under the bed posts. Specifically, let us consider the data reported in the table, compute the coordinates x_c and y_c for each of the weights placed on the bed. Let us assume that the dimensions of the bed are as follows (Fig. 12):

Bed length: $L = 200\,cm = 2.0\,m$ (78.74")
Bed width: $W = 150\,cm = 1.5\,m$ (59.055")

Hint: *Notice that in the case #0 no weight was placed on the bed. You can use this case to obtain the information that you need for to calculate W_{bed} and $x_{c,bed}$.*

Solution See Fig. 13, which shows the solution in the form of a table with attached the load diagrams for the five scenarios.

3.6   Interesting Observations

One would expect that, by adding an extra weight on the bed, all the load cells would record a higher mass. But this is not always the case!

Let us take a second look at the example described in the previous section. In the absence of an extra weight on the bed (see case #0), the load cells in positions 1 and 2 measure 11 kg each, whereas the load cells in positions 3 and 4 measure 9 kg each. Let us now add an extra weight of 1.5 kg on the bed (see cases #1, #3 and #4). In case #1, all the load cells record an increased mass, as we would have expected. However, in case #3 we see that the load cell in position 1 measures $R_1 = 10.375$ kg, which is less than 11 kg! We observe a similar phenomenon in the case #4 with the

CASE	#0	#1	#2	#3	#4
Total weight (Σ) [kg]	40.000	41.500	42.000	41.500	41.500
Total barycenter x_bed [m]	0.750	0.750	0.768	0.723	0.759
Total barycenter y_bed [m]	0.900	0.904	0.857	0.940	0.916
Weight position x_c [m]	-	0.750	1.125	0.000	1.000
Weight position y_c [m]	-	1.000	0.000	2.000	1.333

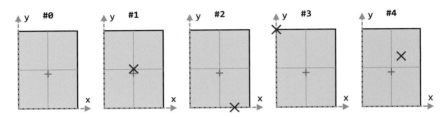

Fig. 13 Table and images showing the solution to the exercise. First the total weight is calculated, then the center of mass of the system (bed + weight) is obtained and, finally, the position of the weight in the four cases is obtained

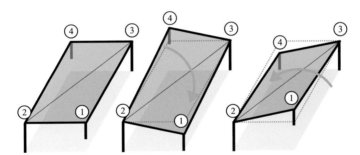

Fig. 14 Representation of the states of equilibrium of a bed that has two longer transverse legs. Left: unstable equilibrium position with the bed parallel to the ground. Middle: stable equilibrium position with the bed tilted to the right; Right: Position of stable equilibrium, with the bed inclined towards the left leg

load cell in the position 2, for which $R_2 = 10.775\,\text{kg} < 11\,\text{kg}$. Interestingly, the extra weight in cases #1, #3 and #4 is the same, but it is placed at different locations with respect to the center of mass of the bed.

This apparently strange behavior can be explained by looking at the following ideal experiment: Let us take a very stiff rectangular board with 4 legs, one at each corner (as a very stiff bed). If two diagonally opposite legs are much longer than the other two (e.g., referring to Fig. 14, assume that legs 2 and 3 are longer), they create a *line of equilibrium*. As the other two legs are shorter, dropping a weight on one triangular side of the board (for instance, putting the weight close to leg 4) leads to a reduction of the reaction force exerted by the leg opposite to the triangular part chosen (in this case a reduction of the weight at the leg 1). This can also bring, in some cases, to the lifting of the bed corner with respect to its original position.

We can, therefore, conclude that using a bed with 4 legs, the equilibrium solution is not necessarily unique, because, for example, by placing the weight exactly on the equilibrium line imagined in the ideal experiment explained above, we can obtain that the table naturally rests more on leg 4 or more on leg 1 based on its initial configuration. But both situations are valid.

On the other hand, considering a triangular table with only 3 legs, we immediately understand that there cannot be any stable oscillatory movement between two legs, and, therefore, each equilibrium solution must be unique. In general we can say that for every structure with more than 3 legs on the ground we have a *hyperstaticity* that generates multiple possible balanced solutions, all equally valid in theory, but only one will appear in practice. This is exactly what Euler studied in his master thesis in 1771!

Challenge Problem 1 Let us now think back at Eqs. (5) and (6). There are many more interesting questions you could explore. Given the same weight, do you think it would be easier or harder to lace it on the bed if the bed's dimensions are closer to a square, in which case $L = W$, or to an elongated rectangle in which case $L < W$ or $L > W$? How do you expect the weight of the object to matter? What are problems that you think you may encounter if the object is very heavy or very light?

3.7   Conclusions

We have seen that it is possible to make measurements of objects placed on a bed using only load cells placed under each post. But what does this have to do with the original question that the doctor asked us? The doctor would have liked to be able to have measurements of cardiovascular health without touching the person. Well, we are now able to say that by analyzing the weight transmitted to the floor by the posts of the bed, we can accurately measure any variation in the center of mass linked to objects placed above the bed. Thus, when a person is on the bed and the blood moves inside the cardiovascular system, we will be able to detect the change of the person's center of mass based on the load cell readings... and here we have our desired BCG!

All of this was made possible by using mathematics to write down the physics of the system into equations. Scientists have taken this approach even further, by using mathematics to write the equations describing the physics of the blood flow into the cardiovascular system and generate virtual BCGs by simulating on a computer healthy or diseased conditions [13].

Examples of real load cell recordings are reported in Fig. 15 These recordings have been performed within the laboratories of the Center for Eldercare and Rehabilitation Technology directed by Prof. Marge Skubic at the University of Missouri. Data have been collected on a healthy subject (male, 26 y/o). By

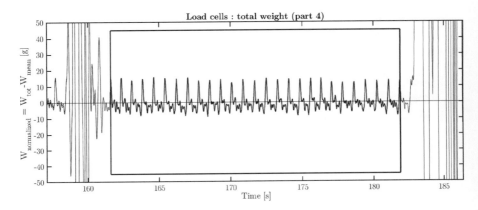

Fig. 15 Total weight signal obtained by summing, for each instant, all four load cells measurements. The result is then normalized by the average

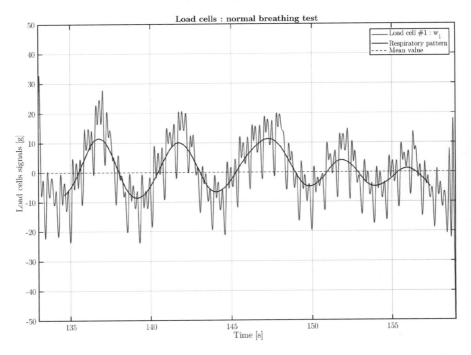

Fig. 16 Over-time weight measurement obtained by one load cell placed under a bed leg with a patient on it. The high-frequency peaks are due to the heart rate, while the low-frequency curve (obtained by interpolation) shows the respiration rate

elaborating the signal from the load cells, it is possible to capture in a very precise manner the heartbeat, see Fig. 15, and the breathing pattern of a subject, see Fig. 16.

Sensors based on the mathematical principles described in this chapter are not just a mere academic exercise. in their different forms, from load cells to accelerometers and piezoelectric sensors, are helping many people around the world suffering from cardiovascular diseases living their lives as independently as possible while being monitored for potential worrisome conditions. Currently, Prof. Guidoboni is directly involved in helping monitor hundreds of elderly individuals in over 10 long-care facilities. Math does, indeed, help us all live a better life

4   Difference Equations in Population Ecology

Section Authors Sergey Lapin, Thomas McCutcheon

Suggested Prerequisites *Algebra, sequences, difference equations*

4.1   Introduction

In many cases it is of interest to model the evolution of some system over time. We can call these systems *dynamic systems*, systems that investigate the quantitative change of variables connected in a system over time. There are two distinct types of dynamic systems. One where time is a continuous variable, and another one where time is a discrete variable. The first case often leads to *differential equations*. The second case can be described using *difference equations*. We will discuss the latter case in this section.

We consider a time period T and observe (or measure) the system states denoted, for instance, by variable U at times $t = nT$, where n is a natural number, i.e., $n \in N_0$. The result is a sequence of states

$$U(0); U(1); U(2); ...; U(k); ...,$$

or using different notation we can write the sequence as

$$U_0; U_1; U_2; ...; U_k; ... \tag{7}$$

The values U_k can obtained from a function f, which is defined for all $t \geq 0$. In this case $U(n) = f(nT)$. This method of obtaining the values is called periodic sampling. We can model the evolution of a system using a difference equation, sometimes it is also called a *recurrence relation*. In this section we will consider the simplest cases first. We start with the following equation,

$$U(n + 1) = aU(n); n \in N_0; \tag{8}$$

where a is a given constant. The solution of (8) is given by

$$U(n) = a^n U(0) \tag{9}$$

The value $U(0)$ is called the initial value. To prove that (9) solves (8), we compute as follows.

$$U(n + 1) = a^{n+1}U(0) = a(a^n U(0)) = aU(n)$$

4.2   Population Growth with Difference Equations

Let us continue our exploration of population growth behavior discussed in Sect. 2. We now investigate how difference equations can be used to track changes in populations.

Understanding how populations grow and what factors influence their growth is important in variety studies such as, for example, bacterial growth, wildlife management, ecology, and harvesting. There are various factors that can affect growth pattern of a specific population, for example, environmental factors, food availability, species interactions, and human interventions. If there is a situation, such that none of these external factors is present to impact the growth and proliferation of organisms, then the organisms will keep on multiplying. Such a growth pattern is called *exponential growth*. Nearly all populations will tend to grow exponentially as long as the resources are available.

Many animals tend to breed only during a short, well-defined, breeding season. It is then natural to think of the population changing from season to season and, therefore, measuring time discretely with positive integers denoting breeding seasons. Growth of such a population can be naturally described using an appropriate difference equation.

Let us start with insect-type populations. Insects often have well-defined annual non-overlapping generations—adults lay eggs in spring/summer and then die. The eggs hatch into larvae which eat and grow and then overwinter in a pupal stage. The adults emerge from the pupae in spring. We take the census of adults in the breeding seasons. It is then natural to describe the population as the sequence of states such as (7), where U_k in this case is the number of adults in the k-th breeding season. The simplest assumption to make is that there is a functional dependence between subsequent generations

$$N_{n+1} = f(N_n), \text{ for } n = 0, 1, \ldots \tag{10}$$

Exercise 5   Let us introduce the number R_0, which is the average number of eggs laid by an adult insect. R_0 is called the *basic reproductive ratio* or the *intrinsic growth rate*. The simplest functional dependence in (10) is

$$N_{n+1} = N_n + R_0 N_n, \text{ for } n = 0, 1, \ldots \tag{11}$$

which describes the situation that the size of the population is determined only by its fertility.

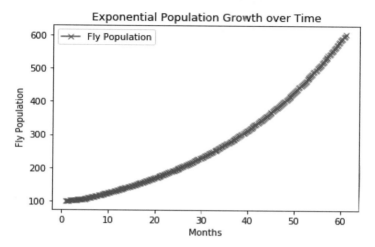

Fig. 17 Exponential population growth

If we determine a population solely by its fertility we get an exponential growth behavior. For instance, Fig. 17 represents the growth of Hessian flies in an ecosystem with abundant food, no predators, and the flies are immortal.

The population will continue to grow exponentially towards infinity which is unrealistic as there is always a finite cap on the resources available in an ecosystem. To this end, the idea of a carrying capacity, often denoted as K, is introduced. K represents the maximum population an ecosystem can sustain.

Exercise 6 We introduce a new functional dependence:

$$N_{n+1} = N_n + R_0 N_n \left(1 - \frac{N_n}{K}\right), \text{ for } n = 0, 1, \ldots \tag{12}$$

In Eq. (12),

$$1 - \frac{N_n}{K}$$

is our carrying capacity term. As the population of our species approaches its carrying capacity the above term approaches zero. This causes our growth term to approach zero which in turn creates what appears to be an asymptote in our graph.

It is important to note that the line created by our carrying capacity term is not truly an asymptote but rather a stable solution. What this mean is that no matter the value of our initial population, the population will always trend towards this line (Fig. 18 and 19).

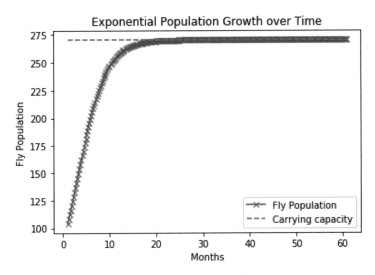

Fig. 18 Exponential population growth with carrying capacity

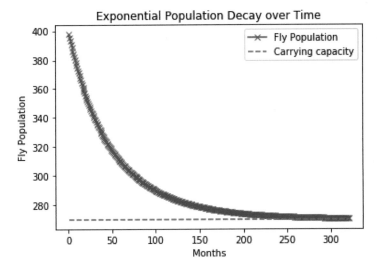

Fig. 19 Exponential population decay with carrying capacity

When a graph acts like Fig. 19 this is called exponential decay. Looking back at Eq. (12), we can see the reason our graph decays is because of the ratio between the current population and the carrying capacity is greater than one. This causes our carrying capacity term to result in a negative number which is then multiplied to our growth term causing the decay we observed. A major difference between stable solutions and asymptotes is that the population graph may cross over our stable solution, however, once it does, as we just discussed, the graph decays back to the stable solution.

4.3   Coding Difference Equations

As nobody wants to perform the thousands of calculations that go into Fig. 21 or any of the other graphs above we outsource those calculations to computer software.

To code the above difference equations we need to create variables representing the current population of each population, separate objects to store all the previous populations for each insect, code the equations and code a loop to run the equations as many times as we need/want.

A question to consider whenever coding is, where might a computers logic cause issues? Most prevalently in our example is population count. To the computer the population count is just a number and if one population goes below zero the computer has no problem continuing to perform the algebra using negative numbers. For example, if the herbivore population was wiped out and ended up as a value below zero to the computer, we would suddenly have a crop population that is growing. How you implement a safeguard to ensure the population count never contains a negative number is up to the coder, however, having a population hit zero can be just as problematic as negative populations. The carrying capacities of the herbivore/predator populations are dependent on the populations of the crop/herbivore which means we may end up dividing by zero if either of those populations crashes down to zero. Most coding environments come with their own error handling and when those environment read divide by zero they halt the code where that occurs and exit. For population difference equation, one population completely wipes out another is very feasible and so we need to design our code in such a way that the error is ignored or can simply never occur. In the case of code used here every divisor that could be zero had a small fraction, such as one over ten million, added so the value there would never truly be zero.

The final topic discussed here is step size. As its name suggests, step size is an incremental increase in a variable, usually x on a standard Cartesian graph like the ones shown above. Step size is often denoted by h and in Eqs. (14), (15), and (16) every term except for the first term, representing the previous population of that organism, would have an h attached to it. The importance of step size in relation to difference equations is best seen in a graphical example. e^x can be approximated by $Y_{n+1} = Y_n + hY_n$ for $Y(0) = Y_0 = 1$. Our approximation comes from a special case of Taylor's theorem. Taylor's theorem allows us to expand any function into a Taylor series provided it is infinitely differentiable at a point a. A Taylor series has an infinite number of terms so we then truncate the series into what is called a Taylor polynomial. In our case e^x is infinitely differentiable for any a contained within the real numbers. All the reader needs to know about derivatives is that e^x is a special case among derivatives and is its own derivative. For more on differential equations and derivatives see Sect. 5. The general form of a Taylor series is $f(x) = f(a) + \frac{f'(a)(x-a)}{1!} + \frac{f''(a)(x-a)}{2!} + \dots$. The apostrophes between f and (a) indicate which derivative of the function is needed, $f'(a)$ is the first derivative of f at a. We are content to use the first two terms of this equation. If $f(x) = e^x$, remembering e^x is its own derivative, our Taylor polynomial becomes $f(x) = f(a) + \frac{f(a)(x-a)}{1!}$. When

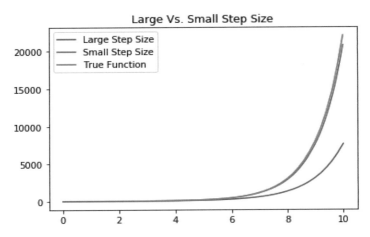

Fig. 20 Large vs small step size

working with difference equations we start with $f(x) = c$ so approximating $f(x)$ is not all that helpful for us. With on slight modification to our Taylor polynomial we can fix that. Let $x = x + h$ and $a = x$. With these changes, our Taylor polynomial is $f(x+h) = f(x) + \frac{f(x)(x+h-x)}{1!} = f(x) + \frac{f(x)(h)}{1!}$. h is our step size and letting $f(x + h) = Y_{n+1}$ and $f(x) = Y_n$ returns us to our stated approximation. As promised we will now look at a graphical example of the importance of step size (Fig. 20).

Clearly a smaller step size produces a graph that is more accurate, however, it is not perfect. Eventually our approximation using the smaller step size will lag behind the true graph. This is because our error, the difference between the true value of the equation and our approximation, is dependent on both step size and, in this case, our function value. For our exponential difference equation the error function is as follows

$$\text{Error} = \sum_{i=2}^{\infty} \frac{h^i}{i!} Y = \frac{h^2}{2} Y + \frac{h^3}{6} Y + \frac{h^4}{24} Y + \frac{h^5}{120} Y + \dots \tag{13}$$

As we can see a large h, a large step size, results in a large error. The h and i terms will be true for any equation while the Y term is specific to our equation. As Y grows our Error term increases and since our approximation is strictly increasing, the further we stretch our range the less accurate our approximation is. As a side note, if you looked at our error terms and said, since those are the difference between the true value of our function and its approximation we should add those terms to our original difference equations, you would succeed at making our approximation more accurate. In this example, with the simple function e^x, we could add more of the error terms onto the equation to get a more accurate approximation. However, calculating and adding these terms can be extremely computationally intensive even for a computer.

If we assume we cannot add more terms to our difference equation to make it more accurate, our only way to decrease our Error is to take smaller and smaller steps. Unfortunately, other errors may arises if we shrink our step size too much. Unlike our previous error, these new errors are the fault of the computer not our equations. We will only discuss round off error here but for those who are interested see Numerical Analysis Mathematics of Scientific Computing by David Kincaid and Ward Cheney [14] for other computational errors. Computers convert numbers into a series of bits, bits are either 1 or 0. However, many bits a computer is using to store a number, it cannot be an infinite number of bits and at some point the computer will round the number [15]. What the round off error of the computer is can vary but if we add too many numbers together the round of error will eventually become significant. For example, let us add one tenth to itself 3.6 million times. if our computer is limited to 24 bit binary storage our binary approximation of one tenth when converted back is off by 0.000000095, this is our round off error. If we then multiply our round of error by 3.6 million, the number of times it has been added together, our error has grown to 0.342. A third may not seem like much, however, this exact error contributed at a catastrophic failure in the Patriot Missile Defense System during the Gulf War [16].

```
PlantPop  =  [];
FlyPop  =  [];
PredatorPop  =  [];

steps  =  300;

DaysinMonth  =  30.44
Hlife  =  44.5
HlifeS  =  (( Hlife / DaysinMonth ) /4)  *steps
Plife  =  76.75
Plifes  =  (( Plife / DaysinMonth ) /4)  *steps

PE  =  (.00001916 *100 )/ Plifes ;  # predator ...
    eat / growth *400 is for scaling to # of months ...
    simulated and our scale being based of 100
HG  =  (0.0001556* 100) / HlifeS ;  # Herbivore ...
    eat / growth
Yc  =  100; # initial Crops
Yh  =  20;  # initial Herbiovre
Yp  =  Yh/1.2; # initial predator
Dh  =  ((44.5/30.44) /(1200000)) =100/HlifeS  %...
    Lifespan of the insect in months scaled to ...
    simulation
```

```
Dp  =  ((76.75/30.44)/(1200000))  *100/Plifes




i  =  1.0;

CropGone  =  0;


for i  =  linspace(1,5,steps)
    Yc  =  max(Yc - (Yh/steps) , 0);
    if(CropGone  ==  1)
        Yc  =  0
    end
    Yh  =  Yh + (Yh * HG  *  (1- max(Yh,0)/(2.7* ...
        Yc + 10 ^ (-10))) - Yh * Dh - Yp * PE);
    Yp  =  Yp + (Yp * PE * ...
        (1-(max(0,Yp)/max(0.000000001,Yh))) - Yp * ...
        Dp);
    if(Yc ~=  0)
        PlantPop  =  [PlantPop,Yc];
    elseif(Yc  ==  0) % ensuring no negative ...
        population
        CropGone  =  1;
        PlantPop  =  [PlantPop,0];
    end
    FlyPop  =  [FlyPop,max(Yh,0)] ;
    PredatorPop  =  [PredatorPop,max(Yp,0)];
end

total  =  linspace(1,5,steps);

for index  =  1:40 % 60 for 150,   20 for 100

    i  =  1.0;

    CropGone  =  0;
    Yc  =  150; % 100 for low density planting

    for i  =  linspace(1,5,steps)  +  4*index
        iNew  =  i - 4*index;
        Yc  =  max(Yc - (Yh /steps) , 0);
```

```
                if (CropGone  = =  1)
                   Yc  =  0
                end
                Yh  =  Yh  +  (Yh  *  HG  *  (1 - ...
                   max (Yh,0) / (2.7*  Yc + 1C ^ ( - 10))) - Yh  *  ...
                   Dh - Yp * PE);
                Yp  =  Yp  +  (Yp  *  PE  *  ...
                   (1 - (max (0 , Yp) / max (0.000000001 , Yh))) - ...
                   Yp  *  Dp);
                if (Yc  ~ =  0)
                   PlantPop  =  [PlantPop , Yc];
                elseif (Yc  = =  0)  % ensuring  no  negative  ...
                   population
                   CropGone  =  1;
                   PlantPop  =  [PlantPop , 0];
                end
          FlyPop  =  [FlyPop , max (Yh,0)]  ;
          PredatorPop  =  [PredatorPop , max (Yp,0)];
          end
          total  =  [total , (linspace (1 , 5 , steps) + 4*index)];

end
figure
plot (total , PlantPop)
title ("Plant  Population  over  Time");
ylabel ("Plant  Population");
xlabel ("Time");
figure
plot (total , FlyPop)
title ("Hessian  Fly  Population  over  Time");
ylabel ("Hessian  Fly  Population");
xlabel ("Time");
figure
plot (total , PredatorPop)
title ("Parasitiod  Wasp  Population  over  Time");
ylabel ("Parasitiod  Wasp   Population");
xlabel ("Time");
```

4.4   Difference Equations for Predator-Prey Problems

Research Project 2 Let us consider a farm with three organisms, wheat, Hessian flies, and Parasitoid wasps. Hessian flies are a pest insect. The adult flies lay eggs on wheat, which hatch into larvae which then devour the stem of the wheat destroying the crop [17, 18]. Parasitoid wasps are a predator insect that lay their eggs into the pupae of Hessian flies, the pupae is the same as a cocoon for caterpillars. The larval wasps then eat the Hessian fly in its pupa stage before progressing to its own pupa stage [19].

In this project we want to see if over planting, which will allow for more herbivores in our system, actually leads to a lower crop yield, due to the increase in herbivore insects. Looking back to our four steps of mathematical modeling, now that we have a goal, to finish step one we must state the assumptions we will be working under:

1. Our farm contains no organisms, including humans, that would affect the population of our crop, herbivore insect and predator insect,
2. We assume all insects alive at the end of one planting season hibernate through to the next planting season,
3. Both of our insects are mono-gendered,
4. We are able to plant 50% more crops in the same space without ruining the soil.

In this second step we will be relying entirely on algebra to create our difference equations. With our assumptions in mind we can now write down our *system* of difference equations. In this second step we will be relying entirely on algebra to create our difference equations.

$$C_{n+1} = C_n - H_n C_d \tag{14}$$

$$H_{n+1} = H_n + H_n G_h \left(1 - \frac{H_n}{r_h C_n} \right) - H_n D_h - P_n G_p \tag{15}$$

$$P_{n+1} = P_n + P_n G_p \left(1 - \frac{P_n}{r_p H_n} \right) - P_n D_p \tag{16}$$

where C_n—Crop population at step n; H_n—Herbivore population at step n; P_n—Predator population at step n; C_d—Crop Destruction Coefficient; G_h—Herbivore growth coefficient; r_h—Ratio of herbivores that can survive per crop; D_h—Herbivore death coefficient; G_p—Predator growth coefficient; r_p—Ratio of predators that can survive per herbivore; D_p—Predator death coefficient.

For the sake of clarity, let us dig into the above variables in more depth. The three population variables can be thought of as the y on a standard graph, for their specific equations. These three are the only true variables as they change with each step in our equations. Everything else is a constant. Before we discuss our constants, we need to touch on the one thing they depend on, step size. Before we begin to use our difference equations we decide our simulation is going to cover a four month period and we are going to take 300 steps during this time. Our step size is thus four divided by 300. We must ensure that all our constants are scaled to this step size otherwise our results will be wildly inaccurate. For example, say a population grows by 10 each month and we take 4 steps per month. If we forget to scale 10 to our step size our population will actually grow by 40. Step size will be discussed in detail in the next section so for now let us return to the discussion of our equation's constants.

Two constants not stated in our equations, but we must consider, are the number of steps in our herbivore's and predator's lifespan. These constants are found by taking the lifespan of the species, usually found in days for insects, dividing by the average number of days in a month then dividing the result by the number of months our simulation spans and finally multiplying by the number of steps we decide to take. Let us call the number of steps in our herbivore's life span S_h and the number of steps in our predator's lifespan S_p. C_d, due to our assumption that no external forces impact the system, is the number of crops one herbivore can destroy over its lifetime divided by S_h. We multiply this to the herbivore population to find the number of crops destroyed during each step. G_h represents how many herbivores are born each step. This constant is found by taking the number of eggs our herbivore species lays in its lifetime converting that to the number of eggs laid per month, multiply by the ratio of average days in a month over the average lifespan in days, and scaling that to S_h. r_h is what it says on the label. If one crop can support 270 of our herbivores eggs r_h is 270. D_h is found by scaling the ratio of the herbivores lifespan in days over the average number of days in a month to S_h. The predator coefficients are found in the exact same manner only scaled to the lifespan of the predator.

The system (14) predicts the size of population total based on the previous population total plus the reproductive growth of the population minus the natural death of our insects minus the consumption of the crop/herbivore by the herbivore/predator.

The graphs in Figs. 21, 22 and 23 were generated under the initial conditions. at time 0, we have 100 crops, 60 herbivore insects and 50 predator insects. In the high density case we have 150 crops initially. We will analyze the behavior of the graphs along with the quantitative, numerical, results to determine if over planting is a viable strategy in our simulation.[1]

It may be tempting to look at the above graphs and state, according to our simulation increasing the density of planted crops will lead to an increased harvest

---

[1] Even though graphs look continuous in time they are composed of many solutions corresponding to discrete time steps.

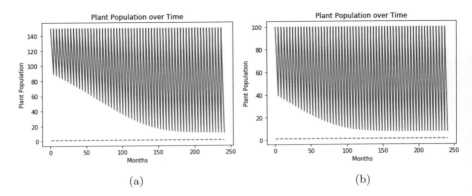

Fig. 21 Plant population. (a) Plant Population in high density scenario. (b) Plant Population in low density scenario

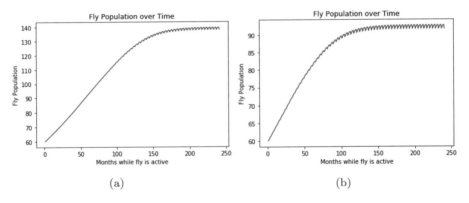

Fig. 22 Prey/pest population. (a) Fly population in high density scenario. (b) Fly population in low density scenario

without the herbivore insect growing out of control. In fact if we assume the farmer goes bankrupt once his crop production falls below a certain level, as they likely would, it seems over planting will keep our farmer afloat. It takes until month 80, which is 20 single planting or 10 biannual planting years, for our farms production to fall to the level of our regularly planted farm. In other words at month 80 our over planting farmer is producing as much our regularly planting farmer produces in their first planting season. However, it is always important to remember the scale of your experiment and what assumptions have been made. A more accurate assessment is that in an isolated system with infinitely nutritious soil the rate at which our herbivore insect reproduces does not increase so rapidly that planting more crops becomes inviable within a reasonable amount of time. Furthermore we can draw more conclusions from our data than just the above. Suppose another swarm of pest infests our farm. The influx of pests has brought us to our critical point, where our pest population graph appears to be approaching the stable solution. We can see that allowing the herbivore insects more resources has increased their stable population, around 137 units of flies. The result of this increase in population is

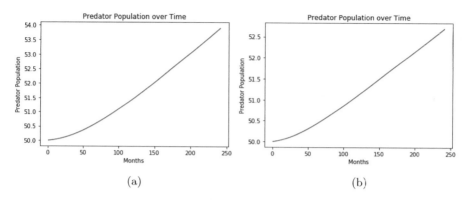

Fig. 23 Predator population. (**a**) Predator population in high density scenario. (**b**) Predator population in low density scenario

that the farm produces 23.49 units of crops per year or about 15.667% of our crops survive. Compare this to our low density case where, at its stable population of 94 units of flies, we yield 16.41% of our planted crop, or about 16.41 units of crop. While our high density crop still yields a greater harvest if we consider the increased costs, both labor and buying the crop, our farmer is likely to earn less if he plants more densely.

A final note about modeling real-world situations is, once your base model is complete, always think about where you could take it next. For example, factoring in soil degradation could give further insight into the potential of denser planting. To take the crop, herbivore, and predator model in a completely different direction one could perform cost benefit analysis using the model to determine at what point our would farmer go bankrupt.

Remark 1 The predator-prey models are not limited to model only behavior of predators and prey (or plants). They are also used in sociology and social psychology. For instance, similar approaches were used in modeling of protest movements [20] or interpersonal relationships [21].

Before we can truly explore our equations we need to learn how to teach a computer to do the leg work and what can go wrong.

5   Modeling the Spread of Infectious Diseases with Differential Equations

Section Authors Alexander Dimitrov and Jacob Pennington

Suggested Prerequisites *Algebra, difference equations, calculus*

This section presents an excerpt of exercises that we use for motivating the modeling aspects of introductory Calculus, typically in a Calculus for Life Sciences class. The two exercises and related labs we discuss here are inspired by material found in the vast repository of modeling scenarios made available by the Systemic Initiative for Modeling Investigations and Opportunities with Differential Equations (SIMIODE) project [22]. The sequences of tasks listed in these labs are not meant to be followed precisely, but to provide some structure to the readers' own investigations, and to illustrate our own thinking around these problems: were we approaching them professionally, we would develop similar lists in order to keep track of what needs to be done, and what we already have finished.

Our entry to Calculus-based modeling starts with something simpler, already discussed in the preceding Sect. 4: a difference equation. We will not repeat the motivation and justification of that section here, but present the scenario leading to a difference equation model. This will allow you to practice that style of modeling some more, and relay it to the broader goal of this section: modeling with *differential* equations, that is, statements relating derivative of functions to hypotheses about properties of the modeled system.

5.1   Modeling the Demise of Candy

The full lab is available as text and code in the MATLAB [5] or Octave [6] environments. We prefer the .mlx Live Script format of MATLAB, since it provides a nice integrated environment in which one can mix text, computations, and visualizations. However, we also provide the .m text-based format which is more versatile and works in both environments.

Exercise 7   You will need M&Ms or Skittles to generate data for the exercise. Dice or coins could also be used with suitable adjustments to the instructions, but they are not as tasty. After we walk through this tutorial together, you will be asked to simulate and model more complicated population dynamics on your own. We recommend completing this Lab 1 in groups of 3–4 students, and having another group of 3–4 similarly minded friends with whom to exchange data.

First, we need to generate some data that represent a declining population or decaying substance. This exercise will be presented as a tutorial. You do not need to complete all iterations, but you should try at least a few tosses to make sure you understand the idea sufficiently well so you can later model the process. Complete the following steps:

1. Start with 40 pieces. Record that number as iteration 0 ($n = 0$).
2. One group member, pick a color of candy at random and keep it secret. Always pick one out of all the colors of candy you had initially, even if no more candies of that color remain.
3. Another group member, shake and throw the pieces onto the table (or onto a paper plate, paper bowl, etc.).

Fig. 24 A graph of the number of remaining M&Ms vs consecutive trial (iteration) number for one possible implementation of the outlined process

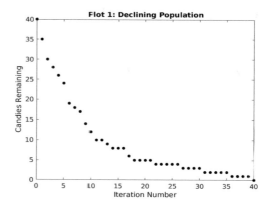

4. Reveal the color chosen in step 2. Remove all candies of that color that have an "M" (or "S") facing up.
5. Record the remaining number of pieces and the iteration number (number of tosses completed so far).
6. If there are no candies remaining, or if you have completed 40 iterations, stop. Otherwise, repeat steps 2 through 5.

Example data for this process is reported as a table in the lab file, generated using M&Ms with five different colors. We present it graphically in Fig. 24. We find that it is always useful to have some visual representation of the process that we are trying to model.

Now let us try to model this process (Step I in Sect. 1.1), in this case as a difference equation. We use an approach that permeates most of the difference and differential equation-based modeling efforts: continuity and conservation. Basically, we assume that our functions do not jump around willy-nilly (are continuous; later we will assume that they are differentiable), and that the quantity that we model does not appear or disappear out of thin air (are conserved), so if we track all the places to which it goes, we can predict the next state of the system. How is that applied to our process of removing M&Ms? Well, we will need some variables first. Let us call the iteration number t (useful for later to indicate *time*), and indicate the number of remaining M&Ms at iteration t as the function $R(t)$. Note that t is an integer here ($t = 0, 1, 2, \ldots$), so formally $R(t)$ is a sequence, usually written in textbooks as R_t. But our notation is also clear, and more useful when we transition to derivatives. Now, assuming we know the value of $R(t)$ ("the current R") and the process by which Rs are obtained, can we say anything useful about $R(t+1)$ (which we read as "the next R")? This is where the idea of conservation comes into place. We strongly suspect that

$$R(t + 1) = R(t) + \text{"stuff added"} - \text{"stuff removed"} \tag{17}$$

In this particular case, there is no "stuff added," so we can set it to zero. But we may have to make some assumptions in other cases, as indicated in Lab 1. "Stuff removed" looks complicated here, with all the random things that we are doing. To simplify, we will ignore the randomness for now (we have other tools to deal with it at a later modeling iteration, in other parts of math—probability), and ask: on the average, how many M&Ms do we remove each time? We realize that we always perform the same process, but in step 4 we will remove more candy if we have more at hand, and less candy if we have less at hand. So we make an educated, and mathematically simple, guess: that we remove a fraction of the available candy; that is, "stuff removed" = $\alpha R(t)$, $0 < \alpha < 1$ (as part of Step 2 in Sect. 1.1). In Lab 1 we assume that $\alpha = 0.1$ (try to figure out why), but it can be a parameter determined later by observations. With that in mind, our "governing equation" for R becomes

$$R(t + 1) = R(t) - 0.1R(t); \quad R(0) = 40,$$

where $R(0)$ represents the initial state of R which is 40 pieces of candy, as per instruction 1.

This equation can be solved to get the explicit expression $R(t) = R(0) * 0.9^t$ (check this on your own, and note further development from Step 2 in Sect. 1.1), but that is not so important, and we cannot get such expressions very often. In Lab 1, we show you how to use the computer to simulate such a system. While writing solutions as expressions is possible for relatively simple systems, the process quickly breaks for more complex system. But the simulations in MATLAB, Octave or other systems are about the same and can produce good results for even complex systems. This simulation-based approach, called "computational thinking," has become more popular due to its flexibility and improvement of technology over the years. First, run a simulation and compare your results to your observations, as suggested in Steps 2 and 3 in Sect. 1.1. Then continue with Lab 1 to try a similar exercise on your own. You will be asked to derive a model and try to both simulate it and analyze it.

In the previous exercise, the data we generated followed a fairly simple pattern: the population/substance represented by the candies always declined/decayed. Now we will complicate the pattern by adding immigrants (or injecting new substance) on each iteration. Continue with the challenge project in Lab 2 on your own, using the techniques that you learned in the previous exercise. See if the results are satisfactory (i.e., represent the observations well), or consider ways to modify your model, completing Step 4 in Sect. 1.1.

Research Project 3 Population Modeling with Skittles
The material in this lab was adapted from exercises available at www.simiode.org, after Brian Winkel, Director SIMIODE, Cornwall NY USA.

(continued)

Problem Statement

In the previous exercise, the data we generated followed a fairly simple pattern: the population/substance represented by the candies always declined/decayed. Now we will complicate the pattern by adding immigrants (or injecting new substance) on each iteration. Remember to reference the exercise material for help with Matlab code and model formulation.

Data Generation

With your group members, pick a number of starting skittles, R_0, between 10 and 20, and a number of immigrants, MI, between 6 and 16. Record these values, but *do not share your immigrant number with other groups*. Then read the following steps, but **do not start the experiment yet.**

(1) Place R_0 candies in one cup (designated the Active cup), and place the remaining candies in a second cup (the Reserve cup).
(2) Shake the Active cup and toss the candies onto the table (or paper plate, bowl, etc.). Remove any candies that do *not* have an "S" or "M" facing up, and place them in the Reserve cup.
(3) Add MI candies from the Reserve cup to the table.
(4) Count and record the number of candies on the table, along with the iteration number (number of tosses completed so far).
(5) Place all of the candies on the table back into the Active cup.
(6) If you have completed 15 iterations, stop. Otherwise, repeat steps 2 through 5.

Note: you can perform more iterations if desired, but 15 is typically sufficient to see a pattern.

Make a Prediction

Before starting the data generation process, answer the following questions:

(1) Describe what you expect the outcome to be. How will the number of remaining candies change as the iterations are carried out?
(2) State your assumptions about the physical activity. Try to break your assumptions down into simple statements: i.e., if you find yourself using "and" a lot, you are probably stating multiple assumptions as one item. Which of these assumptions (if any) influenced your prediction in (1)?
(3) **Now you may proceed with data generation.**
(4) Compare your prediction in (1) with what actually happened.

(continued)

(5) Did any of your assumptions in (2) appear to be proved correct/incorrect? If so, how did they play a role in the experiment?

Report Your Data

(6) Input your data into Matlab (see Table 1 from the tutorial).
(7) Plot your data (see Plot 1 from the tutorial).

Model Development

(8) Describe, in words, what happens to the number of candies remaining after each iteration of the experiment.
(9) Propose a model by translating your description in (8) into a recursive equation.
(10) Justify the reasonableness of your model. Why is each term in the model included?
(11) If your model contains any fixed numbers, try to replace them with model parameters and explain their interpretation. For example, in the tutorial we replaced the value 0.1 with the parameter α to represent the average proportion of candies removed on each toss.

Testing the Model

(12) Generate some data using your model, and plot the model output alongside your data (see Plot 2 from the tutorial). If the model does not appear to match the data well, try adjusting the values of some of your model parameters.
(13) Comment on the success of your model. Does it do a good job of describing your data? Why or why not?
(14) What is the long-term behavior of your model? Does it settle to some fixed value? Does this match the behavior of your data?

Swap Your Data

(15) Exchange your data with another group. Then, use your model to try to estimate the other group's immigration number, MI. *Hint: you will likely need to adjust the value(s) of your model parameter(s).*

5.2   Population Growth Models in Continuous Time

While difference equations are interesting and powerful models for certain natural processes, as outlined in Sect. 4, here we mostly use them to transition to modeling with differential equations. Let us start with Step 2 in Sect. 1.1 again. Previously we interpreted $R(t+1)$ as "the next time we measure R," but we were unclear as to how exactly we measure time, and what exactly is "next." Statements about derivatives try to clarify a bit better what "the next time" means. So, we will assume that time is continuous, represented by some real numbers, possibly with some units attached to it (e.g., seconds). We further assume that "next time" is some predefined small time later than our current time, denoted by δt. Note that in this part of math, δ is usually interpreted/pronounced as "small change of." With this increased precision, we can now model processes like Eq. (17) for populations that do not increase or decrease at specific times, by tracking subsequent states that are closer and closer in time to the current state:

$$R(t + \delta t) = R(t) + \text{"stuff added"} - \text{"stuff removed"} \text{ during time period } \delta t \quad (18)$$

We tend to read this expression as *the system state R a little later (t plus "a little bit")* equals *the system state R now, plus* "stuff added" *minus* "stuff removed" *during the indicated time period.*

   Let us switch our modeling context now from candy to biological population growth models, like the ones in Sect. 4, and reinterpret R to mean "the size of the population of interest." That will allows us to make some hypotheses about "stuff added" and "stuff removed" as part of Step 2 in Sect. 1.1 in the context of biological populations. It is usually easier to reason about "stuff removed": there is some death process, and in the simplest form we can hypothesize that the amount of death is proportional to the population size. In that case, "stuff removed" $= d * R(t) * \delta t$. We also include the assumption typical in this case that death is proportional to time: the longer the interval δt over which we observe the death process, the more individuals will die. Initially we can make similar assumptions about the birth process, or "stuff added." It sounds reasonable to assume that a fraction f of the population would be able to give birth in a small period, and that each birth produces a litter size of l, to obtain "stuff added" $= b * R(t) * \delta t, b = f * l$. Combining the two statements, we obtain

$$R(t + \delta t) = R(t) + (b - d) * R(t) * \delta t \quad (19)$$

   We continue with a few simple algebraic manipulations: subtract R on both sides and divide by δt (now you see why we needed δt in the definition) to obtain a "difference quotient" version of the same equation,

$$\frac{R(t + \delta t) - R(t)}{\delta t} = (b - d) * R(t) \quad (20)$$

As Calculus teaches us, at this stage we can take $\lim_{\delta t \to 0}$ on both sides of the equation to obtain one of the classic population growth models: the proportional or Malthusian growth model [23],

$$\frac{dR}{dt} = R'(t) = k * R(t), \tag{21}$$

where we have substituted $k = (b-d)$. In many introductory books on mathematical modeling this is also referred to as an *exponential growth model*, because its solution can be written as $R(t) = R_0 * e^{k*(t-t_0)}$, a scaled exponent with rate k. As this is not immediately obvious, we avoid that nomenclature here. But we encourage you to continue this development of Step 2 in Sect. 1.1 and (a) verify that the exponential function $R(t)$ is actually a solution to this model, with an initial condition $R(t_0) = R_0$, and (b) learn this mathematical synonym: Eq. (21) with $R(t_0) = R_0$ and $R(t) = R_0 * e^{k*(t-t_0)}$ are equivalent mathematical statements. We do not spend much time showing you how to "solve" such equations, because there are relatively few equations with solutions that can be written as formulae. However, this type of equation (linear autonomous) appears in many approximations and analysis, so it is useful to keep the exponential solution in mind.

The exponential solution and our knowledge of how exponents, and biological systems, behave also suggest that this is *not* a very good model. In particular, the model predicts that, if we start with some known population of size R_0, the population would either explode with exponential growth (if $k = b - d > 0$, birth rate higher than death rate) or decay exponentially to 0 (if $k = b - d < 0$, birth rate lower than death rate). We know of extinct populations, so the latter outcome could be feasible. But we definitely would have noticed if we were knee-deep in bacteria, or fungi, or slime molds, or any other species that has been around for over 3.5 billion years and has had time to grow exponentially. As this is not the case, we will have to modify the model so it better reflects our observations of reality. This observation is a perfect example of the start of Step 4 in Sect. 1.1, at which we evaluate the model and identify shortcomings, leading to alternative models.

Back to the modeling drawing board: improved single population models rely on an observation made by biologists [24, 25]: unconstrained proportional growth is often a good assumption for small populations, but once a population grows large relative to its environment, its growth rate decreases, often approaching 0. We account for this observation by modifying the per-capita growth rate k. In place of a constant, as in Eq. (21), we will use a function of the population size, $k(R)$. As noted in the previous Sect. 4, the simplest function that often produces good results in practice is a linear function, $k(R) = k_0 * (1 - \frac{R}{K})$. This value of this function is close to the constant k_0 when the population size R is small, and close to 0 when the population size is near the environment capacity K ("kapazität" in German, where this model initially came from, hence the default notation). This leads to the differential version of the logistic growth equation (12),

$$R'(t) = k_0 * \left(1 - \frac{R(t)}{K}\right) * R(t) \tag{22}$$

This completes an iteration of Step 4 in Sect. 1.1, leaving us a new model to justify, apply and analyze with Steps 2 and 3 in Sect. 1.1.

This differential equation also has a known formula for a solution, but it is not very helpful for our goals, so we omit it here. Feel free to investigate it on your own. Its dynamics are similar to the discrete logistic model shown in Fig. 18, but in continuous time. More interesting are the local solutions. As we noted, when the initial condition R_0 is close to 0, then Eq. (22) is well approximated by the proportional growth equation (20) and a corresponding exponential growth solution. Correspondingly, if R_0 is close to K, then we can express the population size relative to a small change $u(t)$ around K, $R(t) = K - u(t)$, and obtain an approximate equation for $u(t)$:

$$u'(t) \approx -k_0 * u(t) \tag{23}$$

with a solution which is a decaying exponential with $u(t)$ approaching 0 exponentially, and hence $R(t)$ approaching K exponentially, with a decay rate equal to the negative of the growth rate near 0, k_0. See exercises in Lab 3 for more experience with that approach and note the iterative application of Steps 2 and 3 in Sect. 1.1 here.

5.2.1 A Basic Model of Infectious Disease Spread

We introduce the logistic growth model and approximations so we can use it as a foundation to model the spread of infectious diseases, in particular, the Ebola outbreak in West Africa in 2014 [26]. You can follow the development of the model and preliminary analysis in the supplied Lab 3, with data provided by the World Health Organization (WHO). After completing the lab, try replacing the Ebola data with current data on infections or deaths due to COVID-19 (e.g., in [27, 28]).

How does the spread of Ebola relate to population growth? We provide some empirical, data-based ideas in Lab 3, but we can also think about it here. What population would be modeled? The population of disease-infected people. Then the growth rate becomes intuitively clear: that is the infection rate of the disease. What then is the "carrying capacity" of the environment in this case? What is even "the environment" for the virus which causes the disease? A change in perspective lead us to consider the virus as "the population" that we model, and the people as the substrate/environment that the virus uses to grow. Once we make this conceptual switch, a model for the viral "population" growing in the "environment" of people makes sense and leads us to consider the logistic growth model in this situation as well.

At this point, open Lab 3 in either MATLAB or Octave and follow the instructions and cues there to attempt to model the observations provided by WHO.

We have two challenges that point to the applicability of this model for real-world situations, which follow the text of the project.

Research Project 4 Modeling the Spread of Infectious Diseases
After Lisa Driskell Computer Science, Mathematics, and Statistics Colorado Mesa University Grand Junction CO 81501 USA ldriskel@coloradomesa.edu

Problem Statement
Ebola virus disease or EVD is a potentially fatal disease affecting humans. The virus is thought to be carried by animals such as fruit bats and was discovered in 1976 when the first human outbreaks occurred [1]. The World Health Organization (WHO) recognizes March 2014 to be the start of the worst Ebola outbreak in history. The outbreak was primarily contained to West Africa, beginning in Guinea and spreading to Liberia and Sierra Leone [1]. There was a recent Ebola outbreak in Congo [7]. The outbreak and swift spread of Ebola in West Africa sparked fears worldwide for its potential to reach other regions. In September 2014, the Center for Disease Control (CDC) cited model predictions for the number of Ebola cases that reached as high as 550,000 cases by January 20, 2015 in Liberia and Sierra Leone alone [2]. When the model was corrected for under reporting, the number of cases jumped to 1.4 million. At the time of the prediction, Liberia and Sierra Leone combined for 2407 cases while the model fairly accurately estimated 2618 cases [2].

In a January 2016 news release, the World Health Organization officially declared the Ebola outbreak over for all three countries [3] (although few flare-ups resulted in 14 additional cases since that declaration). Because we have the data for the actual epidemic, we now know that the predictions by the CDC vastly overestimated the impact of the outbreak. We will investigate our own models of the epidemic based on the data that we now have.

We will use data published by the WHO ([4, 5]) to determine parameters for simple models that describe the outbreak. The data for the total number of Ebola cases includes cases confirmed by test results as well as suspected cases (those showing symptoms resistant to treatment) and probable cases (suspected cases that have been seen by a clinician). Also note that the data provided for the total number of cases is a running total of the number of cases which includes those patients who may have since recovered or died from the disease. The data for the number of deaths due to Ebola is also included to provide the opportunity for additional modeling and comparisons.

(continued)

References

[1] World Health Organization. 2016. Ebola virus disease. World Health Organization (Fact sheet No.103). Available at http://www.who.int/mediacentre/factsheets/fs103/en/. Accessed January 2016.

[2] Meltzer, M. I., Atkins, C. Y., Santibanez, S., et al. 2014. Estimating the Future Number of Cases in the Ebola Epidemic -Liberia and Sierra Leone, 2014-2015. Centers for Disease Control and Prevention (MMWR; 63,Suppl-3). Available at http://www.cdc.gov/mmwr/preview/mmwrhtml/su6303a1.htm. Accessed January 2016.

[3] World Health Organization. 2016. Latest Ebola outbreak over in Liberia; West Africa is at zero, but new flare-ups are likely to occur. World Health Organization (News Release; January 14, 2016). Available at http://www.who.int/mediacentre/news/releases/2016/ebola-zero-liberia/en/. Accessed January 2016.

[4] World Health Organization. 2014. Ebola virus disease. World Health Organization (Disease Outbreak News; March 22, 2014-August 31, 2014). Available at http://www.who.int/csr/don/archive/disease/ebola/en/.

[5] World Health Organization. 2014-2016. Ebola virus disease. World Health Organization (Situation Reports; September 5, 2014-June 10, 2016). Available at http://www.who.int/csr/disease/ebola/situation-reports/archive/en/.

[6] World Health Organization. 2016. Ebola virus disease. World Health Organization (Situation Summary; April 6, 2016 -May 11, 2016). Available at http://apps.who.int/gho/data/view.ebola-sitrep.ebola-summary-latest. Accessed Sept 2018.

[7] 2018 Kivu Ebola Outbreak. https://en.wikipedia.org/wiki/2018_Kivu_Ebola_outbreak. Accessed Jan 28, 2019.

Heads Up
Unlike the previous lab, the **CHALLENGES for this project are mixed in throughout**. Make sure you do not miss any!

We continue with some guided exercises, and challenges, which will help us approach this project.

Loading The Data
First, make sure to download the file *eboladata.xlsx* and save it in the same directory as this *Matlab* script. Then the data can be loaded using the following code.

```
table  =  readtable ("eboladata.xlsx", ...
    'PreserveVariableNames', true);
data  =  table2array(table(:,1:2));  % convert to ...
    array for easier plotting
day  =  data(:, 1);
cases  =  data(:, 2);
```

Now the *day* and *cases* variables store vectors representing the number of days since reporting began and the total number of cases reported, respectively.

Data Visualization

Next, let us plot the data to get an idea of what the course of the epidemic looked like.

```
figure('Name', 'Plot 1');
plot(day, cases, 'k.', 'MarkerSize', 10)
title('Full Dataset')
xlabel('Day');
ylabel('Total Cases');
```

The data follows a clear pattern: cases increase slowly at first, then much more quickly, and then increase very slowly again. This aligns with a typical trend in epidemics: at first the disease spreads slowly from a handful of carriers, then spreads rapidly as the number of carriers becomes large, and finally slows down as the outbreak is controlled through vaccination, social distancing, and other public health measures. However, it is not immediately clear how we should model this trend. So instead of tackling the full dataset at once, we will try breaking the data (and our model) into pieces.

Let us look at just the first 200 days of the outbreak. We can use the *find( )* function to identify the relevant entries in the *day* variable

```
day200  =  find(day <= 200);
```

The new variable *day200* can now be used as a shortcut to select the relevant datapoints from the *day* and *cases* vectors.

Challenge Problem 2

(1) Plot the total number of cases for the first 200 days of the outbreak.
(2) Describe the trend in the data for this portion of the outbreak.

Model Development

As with the previous lab, while we could try to directly model the outbreak in terms of total cases over time, we can make our modeling task easier by instead focusing on incremental changes. Look back at your plot from exercise (1): how quickly is the number of cases changing when the number of cases is small? How quickly is the number changing when the number of cases is large? You should see that the rate of increase is smaller for fewer cases and greater for more cases. In other words, *the rate of change in the number of cases is proportional to the number of cases*. We can express this observation as a differential equation:

$$dN/dt = rN, N(0) = N_0$$

where N represents the number of cases, r is a constant that determines the magnitude of the relationship and N_0 is the population size at the initial time (initial condition). Notice the similarity between this model and the model we came up with in the first part of the previous lab. In that case, we represented the decline of the skittles with a recursive equation:

$$R(t + 1) = (1 - a)F(t)$$

where $R(t)$ represents the number of skittles remaining after n tosses. Both of these models make a similar statement: that some variable is changing in proportion to the current value of that variable. However, in the skittles exercise, we were modeling a *discrete* process: the number of skittles could only change on each toss, with no concept of "fractional tosses." For the Ebola outbreak, our data is *continuous*: the true number of cases might change every day, or every hour, or every second (even though the timing of the reports will be at discrete intervals).

Now, we would like our model in a more convenient format: in short, we want to be able to ask "how many cases will there be on day t?" With a differential equation, we cannot answer this question immediately: the equation only tells us about rates of change. However, if we know (or at least be able to guess) the state of the equation at some initial time, and the value of any parameters in the equation, we can obtain a solution as an expression for the function $N(t)$, or approximate the solution with numerical simulations. In this case we will do both. In most cases we cannot write an *expression* for that solution. That is why we use simulations, but validate them when we can against known solutions.

Remember, $n0$ is the initial number of cases. The parameter r controls how quickly the number of cases increases (or decreases). Let us try to estimate both parameter values from the data.

Challenge Problem 3

(3) Use the data to make an initial guess for the value of the parameter $n0$, then enter it in the code below.

(4) Use the data to make an initial guess for the value of the parameter r, then enter it in the code below.

Hint: solve the equation $N(t) = N_0 * e^{-k*t}$ for r.

```
initial_N0  =  1;        % (answer to 3 goes here )
initial_r  =  1;         % (answer to 4 goes here )
```

After entering your answers, rerun the code for this section.

Testing the Model

Now that we have some parameter values to start from, we can plot the model against our data. The following code will define the model as a function of the number of days since the outbreak, t, and produce the desired plot.

(5) Run the code block below to plot the comparison. Then, enter new values for the parameters $n0$ and r and rerun the code block to see the new comparison. Try to find values of $n0$ and r that make the model and data match as closely as possible.

```
N0  =  initial_N0;   % updated parameter values ...
       go here
r  =  initial_r;     % (you may need to adjust ...
       one or both of these values)

% define the known solution function
Nsol  =  @(t) N0*exp(r*t)

% alternatively , simulate the system for some ...
       time points
tspan  =  [0 200];   % interval of time to use
[t , Nsim]  =  ode23(@(t ,N) r*N, tspan , N0);   % ...
       simulates the model , stored as vectors t and N
% note how ode23 has to be formatted to work , ...
       and what kind of outputs it
% produces by checking the documentation. ...
       There are also ode45 , ode23s and
% a few other simulators with various properties

figure ('Name' , 'Plot 2');
```

```
plot (day (day200),  cases (day200),  'k.',  ...
    'MarkerSize',  10)  % the data
hold on
fplot (Nsol,  tspan,  'b-')  % exact solution
plot (t,  Nsim,  'ro')  %  model simulation
ylim ([0,max (cases (day200))*1.1])  % restrict ...
    the vertical axis to known data ranges
title ('Modeling the Early Outbreak')
xlabel ('Day');
ylabel ('Total Cases');
hold off
```

Assuming you found a suitable combination of parameter values, you should see that the model does indeed track this portion of the data fairly well. However, so far we have only compared the model to the first 200 days. What about the rest of the data? The following code defines some new variables for selecting the first 300 days instead of the first 200 and regenerates the model on this extended set of points.

```
day300  =  find (day < = 300);
tspan  =  day (day300);
[t, N]  =  ode23 (@(t,N) r*N, tspan, N0);
```

(6) Generate a new plot that compares the model to the first 300 days of data. You can use the plotting code from exercise (5) as a template if needed. You may also change the number of days shown if desired (you could try stopping at 250 days or 400 days, for example).

(7) You should see that the model does not work very well beyond 200 days. Describe any differences you see. What happens in the data as t becomes larger? What happens in the model as t becomes larger?

Model Development

Clearly we need to make some adjustments if we want our model to describe the full course of the outbreak. Let us look at the full dataset again to remind ourselves what the trend looks like.

```
figure ('Name',  'Plot 3');
plot (day,  cases,  'k.',  'MarkerSize',  10)
title ('Full Dataset')
xlabel ('Day');
ylabel ('Total Cases');
```

The key difference appears to be that, while the number of cases in our model continues to accelerate as the number of days increases, the number of cases in the data begins to decelerate somewhere around 250 days and eventually stops changing much at all. So, we need to modify our model in a way that causes the rate of change in the number of cases to become small as the number of cases approaches some upper bound. Let us state this observation in a new differential equation:

$$dN/dt = rN * f(N, K)$$

where $f(N, K)$ represents some function of the number of cases and the upper bound that the data approaches—often this latter parameter is referred to as the *carrying capacity*. But what should the function $f(N, K)$ look like? We would like $f(N, K)$ to be close to 1 when the number of cases is small—that way, the value of $r N$ will be close to what it was for our previous model, which worked pretty well when the number of cases was small. On the other hand, we want $f(N, K)$ to be close to zero if the number of cases is getting close to the upper bound, K. That way, the rate of change will approach 0 as the number of cases increases, similar to our data. One form of $f(N, K)$ that satisfies these properties is:

$$f(N, K) = 1 - N/K$$

making our differential equation:

$$dN/dt = rN(1 - N/K)$$

Before we go on, let us convince ourselves that this formula does in fact satisfy the properties we just mentioned.

Challenge Problem 4

(8) What is the value of $r N (1 - N/K)$ if N is small, say $N = 0$?
(9) What is the value of $r N (1 - N/K)$ if N is very close to K, say $N = K$?
(10) (Optional) Can you think of any other version of $f(N, K)$ that might work? Try some out!

Testing the Model

Now that we are satisfied with the reasonableness of our new model, let us simulate it. This model also has a known expression for a solution, but we will proceed with simulation-only approach in order to prepare you for more complex cases. As before, we need to assign some initial guesses for our model's parameter values. To prevent overwriting the variables we used before, we can name our new parameters *n02, r2*, etc. Fortunately, since our new model is very similar to the old model for certain portions of the data, the values we already determined for *n0* and *r* should be decent guesses for *n02* and *r2*.

```
initial_N02  =  N0;
initial_r2  =  r;
```

This just leaves the parameter *K*.

Challenges

(11) Use the data to make an initial guess for the value of the parameter K, then enter it in the code below. Recall that *K* represents the upper bound for the number of cases.

```
initial_K  =  10000;   %  (answer to  11  goes ...
      here  )
```

After entering your answer, rerun the code for this section.

Now we can compare the model to the data as before.

(12) Run the code block below to plot the comparison. Then, enter new values for the parameters *n02, r2*, and *K*, and rerun the code block to see the new comparison. Try to find a combination of parameter values that makes the model and data match as closely as possible.

```
N02  =  initial_N02;   %  Updated  parameter ...
      values  go  here.
r2  =  initial_r2;   %  You  may  need  to  change ...
      one
K  =  initial_K;   %  or  all  three  of  these ...
      values.

tspan2  =  [day(1)  day(end)];
[t2,  N2]  =  ode23(@(t2,N2)  r2*N2*(1-N2/K), ...
      tspan2,  N02);

figure('Name', 'Plot 4')
plot(day,  cases,  'k.',  'MarkerSize',  10)
hold  on
plot(t2,  N2,  'ro')
ylim([0  max(cases)*1.1])   %  again  restrict ...
      vertical  axis  to  range  of  observed  cases
```

```
title ('Modeling the Full Outbreak')
xlabel ('Day');
ylabel ('Total Cases');
hold off
```

(13) Would you say the model does a good job of describing the Ebola outbreak? Why or why not?
(14) List one difference you notice between the data and the model.
(15) Propose an explanation for the difference you noted in (14), in terms of the epidemic. You could also think of this in terms of what pieces the model might be missing.

5.3 Discussion

Back to our project-based challenges. The first attempt was to use the approximate proportional/exponential growth model based on early observations and try to predict the course of the disease. When you compare to the actual course, you will see that the model works well initially, but then drastically over-estimates the infections. That leads us to search for an alternative, and the logistic model provides a reasonable option. Once we have a reasonable model like the logistic growth model, we can try to assess key features of the outbreak process by analyzing the model itself. If you look at a result from Lab 3 that we show on Fig. 25, you will notice 2 things: first, the model matches observations reasonably well initially, but diverges significantly in later days. A logistic model cannot do much better than that, since it has some symmetries which the real data does not, so we will need to further revise the model to improve our prediction (Step 4 in Sect. 1.1).

Second, and perhaps more interesting, we see one prediction of the model that is mirrored in the data: the functions describing the total infected population size change their nature, from increasing initially, to decreasing in the later stages of the infection. We can translate that observation in mathematical statements, as per Step 2 in Sect. 1.1. In Calculus, such a transition from increasing to decreasing (or vice versa) happens at *inflection points*. Since this is obviously an important point to know for the epidemic, both the time at which it happens and the size of the infected population at that time, let us try to analyze the equation in order to find the inflection point. In principle, we can use the simulations in Lab 3, and you should try that, for some more "computational thinking." But we can also use analytical results based directly on the equation, without simulations or a known formula for a solution. One thing to note: when we have an expression about the derivative of a function, as in (22), we know *a lot* about the function itself. We know its first derivative, which means we can take a derivative and compute its second derivative. Then one more time, to obtain its third derivative, etc. So, basically, from the Eq. (22) we can find out ALL derivatives of R and only do not know the function itself, values for the population size R at specific times t!

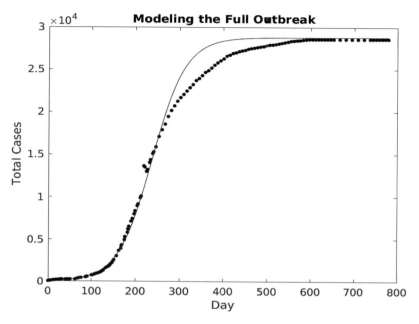

Fig. 25 A graph of the WHO Ebola outbreak data (black dots) and a logistic model fit to it (red curve). We see that the model performs well initially, but has problems explaining the data later

Let us work out this analysis here. How can we find an inflection point of a function? Well, if the function transitions from increasing to decreasing, that means there is a minimum of the derivative of the function, so that the derivative is positive to the left of it (increasing) and negative to the right (decreasing). This means that an the inflection point the derivative of the derivative will be zero, which means that the second derivative of the solution is zero at an inflection point. Now that we have a feature to look for, we take the formula for the solution, find the second derivative, ... Oh, wait, we *do not* have the formula of the solution! However, we do have a formula for its derivative that was our model expressed as a differential equation (22)

$$R'(t) = k_0 * \left(1 - \frac{R(t)}{K}\right) * R(t)$$

This is actually nice, since we do not have to compute the first derivative, we already know it. So if we take one more derivative of this equation, we will get

$$R''(t) = \frac{d}{dt}\left(k_0 * \left(1 - \frac{R(t)}{K}\right) * R(t)\right)$$

Hmm, we have the second derivative, but the right hand side seems a bit messy. Do not we need $R(t)$ after all? Let us persist, and evaluate the derivatives while keeping the unknown function $R(t)$ and its derivative $R'(t)$ as we come across them. Continuing with the product rule of the right hand side, we obtain

$$R''(t) = k_0 * \left(-\frac{R'(t)}{K} * R(t) + \left(1 - \frac{R(t)}{K} \right) * R'(t) \right) = k_0 * R'(t) \left(1 - 2\frac{R(t)}{K} \right)$$

That does not look too enlightening, until we remember that we **do** have an expression for $R'(t)$—that is the original equation (22). We substitute this expression for $R'(t)$ on the right hand side to obtain

$$R''(t) = k_0 * k_0 * \left(1 - \frac{R(t)}{K} \right) * R(t) * \left(1 - 2\frac{R(t)}{K} \right)$$

That equation is almost ready for analysis. For what values of $R(t)$ would $R''(t)$ be zero? From the right hand side, we see that $R(t) = 0, K$, or $K/2$ are all possible solutions. However, $R(t) = 0$ and $R(t) = K$ represent "no infection" and "everyone infected" states, respectively (known as *fixed points*), so we are left with the conclusion that the inflection point occurs when $R(t) = K/2$, or half of the infectable population is infected. This analysis does not produce the time at which this inflection occurs: for that information, we would still need a formula or simulation. Try to find the time of inflection with a simulation in Lab 3. However, this approach *does* provide two useful predictions. First, it predicts a level of infection beyond which things would get better. Second, it predicts that, according to this model, it will take the same amount of time for things to get better as it took to get to that situation. Those are powerful and actionable predictions, and similar models have been used by WHO to respond to the Ebola, COVID, and other, epidemics.

6   Conclusions

In this chapter we presented a progression of biological situations, coupled by associated mathematical representations that reflect our understanding of these real-world situations. This is the gist of mathematical modeling. It is a process of discovery, and it is an iterative process, as summarized by Steps 1 through 4 in Sect. 1.1. The most important of them is probably Step 4 in Sect. 1.1, which asserts that anything we write down is not "the truth," but just our latest, most improved understanding of the real-world situation that we are translating into a mathematical representation. We are using the term "translate" quite literally here, since we think of mathematics as a language (or rather, a set of related languages), but a language which can provide much higher precision of expression compared to other human languages.

We as professional applied mathematicians are directly involved with using very similar models to approach and solve real-world problems. As noted previously, Prof. Guidoboni is currently directly involved in helping monitor hundreds of elderly individuals in over 10 long-care facilities. The predator-prey models discussed in Sect. 4 are not limited to model only behavior of predators and prey. They are also used in sociology and social psychology [20, 21]. Students of Dr. Dimitrov's used a version of the Ebola modeling project as a stepping stone to identify and model problems from the current COVID pandemic and ultimately analyze different approaches to hospital utilization to help states manage the influx of COVID patients [29]. Should you chose to proceed on this path, soon you will be able to tackle difficult societal problems and help people on your own, through mathematical modeling.

We tried to demonstrate the breadth of situations that can be understood using mathematical models, while remaining within the domain of biological systems. Even that shows the breadth of problems that can be addressed mathematically, and the scope of results and conclusions that can be obtained in that way. You will hear this statement many times in your journey building and evaluating mathematical models, but it bears repeating: just start, summarize your thought in mathematical notation, and then let the math help you on the way.

References

1. G. Polya. *How to solve it.* Princeton University Press, 1945.
2. K. K. Bliss, F. K. Fowler, and B. J. Galluzzo. *Math Modeling: Getting Started and Getting Solutions.* SIAM, Philadelphia, 2014.
3. S. Buyting. Women and science suffer when medical research doesn't study females. https://www.cbc.ca/radio/quirks/july-25-2020-women-in-science-special-how-science-has-done-women-wrong-1.5291077/women-and-science-suffer-when-medical-research-doesn-t-study-females-1.5291080, 2019. Accessed: 2021-02-07.
4. Code files for the chapter. https://github.com/adimitr/ModelingBioChapter. Accessed: 2021-06-30.
5. Matlab:. https://www.mathworks.com/products/matlab.html. Accessed: 2021-01-30.
6. GNU Octave:. https://www.gnu.org/software/octave/index. Accessed: 2021-01-30.
7. Python:. https://www.python.org/. Accessed: 2021-01-31.
8. Jupyter notebook:. https://jupyter.org/. Accessed: 2021-01-31.
9. Matlab onramp:. https://www.mathworks.com/learn/tutorials/matlab-onramp.html. Accessed: 2021-01-31.
10. Octave tutorials:. https://wiki.octave.org/Video_tutorials. Accessed: 2021-01-31.
11. S Freeman, SL Eddy, M McDonough, M Smith, N Okoroafora, H Jordt, and MP Wenderoth. Active learning increases student performance in science, engineering, and mathematics. *Proceedings of the National Academy of Sciences*, 11(3):8410–8415, 2014.
12. I Starr and A Noordergraaf. *Ballistocardiography in cardiovascular research: Physical aspects of the circulation in health and disease.* Lippincott, 1967.
13. G Guidoboni, L Sala, M Enayati, R Sacco, M Szopos, J Keller, M Popescu, L Despins, V Huxley, and M Skubic. Cardiovascular function and ballistocardiogram: a relationship interpreted via mathematical modeling. *IEEE Transactions on Biomedical Engineering*, 66(10):2906–2917, 2019.
14. D Kincaid and W Cheney. *Numerical Analysis: Mathematics of Scientific Computing.* American Mathematical Society, 2002.
15. JR Hanly and EB Koffman. *Problem solving and program design in C.* Pearson, New Jersey, 2018.

16. DN Arnold. The patriot missile failure. http://www-users.math.umn.edu/~arnold//disasters/patriot.html, 2000. Accessed: 2021-02-02.

17. RB Schmid, A Knutson, KL Giles, and BP Mcornack. Hessian fly (Diptera: Cecidomyiidae) biology and management in wheat. *Journal of Integrated Pest Management*, 9(1), 2018.

18. Roe K. Tolerance as a novel mechanism of hessian fly control on wheat. Master's thesis, Purdue University, 2016.

19. CM Packard. Life histories and methods of rearing hessian-fly parasites. *Journal of Agricultural Research*, 6(10):367–381, 1916.

20. P Oliver and D Myers. The coevolution of social movements. *Mobilization*, 8:1–25, 2003.

21. D Felmlee and D Greenberg. A dynamic systems model of dyadic interaction. *Journal of Mathematical Sociology*, 23(3):155–180, 1999.

22. SIMIODE modeling scenarios. https://www.simiode.org/resources/modelingscenarios. Accessed: 2021-01-30.

23. T.R. Malthus. *An Essay on the Principle of Population As It Affects the Future Improvement of Society, with Remarks on the Speculations of Mr. Goodwin, M. Condorcet and Other Writers*. J. Johnson in St. Paul's Churchyard, London, 1798. https://archive.org/details/essayonprincipl00malt, Accessed: 2021-01-30.

24. P.-F. Verhulst. Recherches mathématiques sur la loi d'accroissement de la population. *Nouveaux Mémoires de l'Académie Royale des Sciences et Belles-Lettres de Bruxelles*, 18:1–38, 1845.

25. McKendrick A. G. and Pai M. K. XLV.—the rate of multiplication of micro-organisms: A mathematical study. *Proceedings of the Royal Society of Edinburgh*, 31:649–653, 1912.

26. World Health Organization. Ebola virus disease. http://www.who.int/mediacentre/factsheets/fs103/en/. Accessed: 2021-02-05.

27. The covid tracking project:. https://covidtracking.com/. Accessed: 2021-02-05.

28. Ensheng Dong, Hongru Du, and Lauren Gardner. An interactive web-based dashboard to track covid-19 in real time. *The Lancet Infectious Diseases*, 20:533–534, 5 2020.

29. M. D. Roberts, H. D. Seymour, and A. Dimitrov. Increasing number of hospital beds has inconsistent effects on delaying bed shortages due to COVID-19. *SIAM Undergraduate Research Online*, 2021.

Population Dynamics of Infectious Diseases

Glenn Ledder and Michelle Homp

Abstract

We develop two models for the spread of an epidemic: an individual-based model and a continuous-time SEIR model. The presentation is accessible to students who have a strong background in algebra and functions, with or without calculus. Projects focus primarily on extending the models to novel situations, including more complicated disease histories and incorporation of mitigating strategies such as isolation and vaccination.

Suggested Prerequisites *Mathematical epidemiology requires strong algebra skills and a good understanding of the function concept. For maximum flexibility, each exercise, challenge problem, and research project is marked to indicate how much background is required in both calculus and computer programming. In both cases, "level 0" means there is no requirement. Level 1 calculus means that students need to understand the derivative concept, which is presented in Sect. 3, while level 2 calculus indicates a requirement for some computation of derivatives and or anti-derivatives. Level 1 programming means that students need to be able to run the programs presented in the appendix using either MATLAB or Octave and also make minimal modifications to commands that assign values to variables. Level 2 programming means that students will need to make some modifications to the program codes in order to adapt the program to a novel scenario.*

1 Mathematical Models in Epidemiology

The reader may be familiar with the basic model of exponential growth:

G. Ledder (✉) · M. Homp
Department of Mathematics, University of Nebraska-Lincoln, Lincoln, NE, USA
e-mail: gledder@unl.edu; mhomp@unl.edu

$$P(t) = P_0 e^{kt}, \tag{1}$$

where $P(t)$ is the population of some group at time t, $P(0) = P_0$ is the initial population, and $k > 0$ is a parameter called the *rate constant* or *proportionality constant*.[1] From this example, we can tease out some details about models and modeling.

1. If you take $P_0 = 1$ and $k = 1$, the population at time 1 is $P = e \approx 2.718$, which is not an integer.
2. If you keep using larger and larger values of t, there is no limit to how large P can be.
3. The model can be rewritten in a different form by taking the logarithm of both sides to get $\ln P = \ln P_0 + kt$. This means that the graph of $\ln P$ vs t is a straight line and suggests a way to identify k from data.

Each of these statements illustrates a typical characteristic of mathematical models. First, models often give results that include all possible numbers in a range on a number line even though the quantities they represent can only take discrete values. While a value of 1079.6 is not technically correct for an integer population, there is no problem interpreting it as approximately 1080. Second, models can sometimes have qualitative properties that are not in alignment with the context they are intended to represent. Third, models can sometimes be rearranged or analyzed in ways that allow for convenient comparison with data or suggest theoretical insights.

While most of us lack the capability to collect data for growth of bacteria or other rapidly growing organisms, those who have done so will tell us that the results are never exactly the same two times in a row. The real biological world has unavoidable randomness that makes results of individual experiments unpredictable.

Although the exponential growth model is not "correct" or "true," it still has value. Data from population growth experiments can adhere closely to the prediction of exponential growth until resource limits start to matter. The extreme example of this is the amount of data that can be put on a state-of-the-art computer chip, a quantity that has been growing exponentially since the 1960s and is only just starting to show signs of slowing [17].

Our epidemiological models are not going to be "true" either. They will be based on assumptions that are oversimplifications. Nevertheless they will have value because they can make general predictions about the course of an epidemic and the impact of public health policies. For example, a model can be used to address questions such as "Will extensive contact tracing significantly decrease the negative impact of an epidemic?" Contact tracing requires significant infrastructure, so we would like to have some idea of whether it will matter before we decide to do it.

---

[1] You may be familiar with this model using different symbols for some of the quantities, such as $y = Ae^{kt}$. The symbols *represent* the quantities, so the model is the same no matter what symbols are used.

Models may not give us exact answers to questions such as these, but they are the best way we have to predict outcomes when we cannot do real experiments.[2]

Based on these considerations, we can make a tentative definition of the term *mathematical model*:

Definition 1 A **mathematical model** is a collection of one or more variables together with enough mathematical equations or rules to prescribe the values of those variables. Models are based on some actual or hypothetical real-world scenario and created in the hope that they will capture enough of the features of that scenario to be useful for answering research questions.

The phrase "enough of the features" requires interpretation. We can ignore minor deviations such as fractional populations or small percentage errors, but we should be on the lookout for qualitative errors. As an example, the model most often used in ecology to describe simple food web systems with one prey species and one predator species is the *Lotka–Volterra* model (see the Wikipedia article [16], for example). We can add an additional term that directly decreases the predator population through hunting. However, the model is a system of interacting components. Mathematical analysis shows that the increased predator death rate is accompanied by an increase in prey population, which in turn causes an increase in the predator birth rate. In the Lotka–Volterra model, this indirect increase fully compensates for the additional deaths, leading to the faulty conclusion that increasing the predator death rate does not actually reduce the predator population. Predator extinction from hunting is widely observed in ecology; the correct conclusion to draw from the model analysis is not that predators do not go extinct, but that the Lotka–Volterra model is fundamentally flawed [10]. Nevertheless, most treatments of the Lotka–Volterra model in books and other references fail to mention this critical flaw.[3]

Plan of the Chapter
Section 2 is devoted to an *individual-based model*, a model based on tracking the characteristics of individuals over time, much as occurs in a computer game incorporating characters with changing ability ratings and who maintain inventories of personal items.

The remainder of the chapter is devoted to *dynamical system models* that use mathematical formulas to update the total counts of people in different epidemiological states, such as *infectious* and *recovered*. The background for these models is presented Sect. 3. A broad classification scheme is presented in Sect. 4 and then the

---

[2] A commonly quoted statement is that "All models are wrong, but some are useful." This statement points in the right direction, but the word "wrong" implies that the purpose of a model is to represent reality. The purpose of a model is to help us obtain insights about reality, so models can be "good" or "bad," but not "right" or "wrong."

[3] While the exponential growth model also fails to predict long-term qualitative behavior, it is useful in the short term. The Lotka–Volterra model is often qualitatively inaccurate in the short term as well as the long term.

standard SEIR example is developed in Sects. 5 and 6. Sections 7 and 8 discuss the
various ways to obtain results from the model.

Exercises appear where needed. There are a small number of challenge problems,
one that asks students to implement a model in a spreadsheet and several that ask
students with adequate calculus background to derive some of the important results
that appear in Sects. 6 and 7. Projects appear at the end of three sections: projects
using individual-based models are in Sect. 2, while projects using dynamical system
models are in Sects. 7 and 8.

All of the exercises, challenge problems, and projects are classified according to
the level of calculus background required and the level of programming challenge.
For both, level 0 means no requirements. In calculus, level 1 means that students
need the background provided in Sect. 3, while level 2 means that calculus
computation is required. In programming, level 1 means that students need to be
able to run MATLAB programs (provided in the appendix) in either MATLAB or
Octave [12] and make minor changes to a few lines of code, while level 2 means
that students have to make modest changes to the program code.

This chapter can be used in different ways, depending on the background of the
students and the amount of time available for background material. Instructors who
want to get to research with a minimal amount of preparation or whose students
have not had calculus can focus exclusively on Sect. 2, which has ample scope for
research projects. Instructors whose students have taken calculus may want to skip
Sect. 2 so as to focus on the more commonly used continuous dynamical system
models. Sections 7 and 8 provide complementary launch points for research, but
either can be done without the other. Figure 1 shows these relationships among the
sections, with one caveat: readers who wish to skip Sect. 7 should still read the short
introduction to that section before proceeding to Sect. 8.

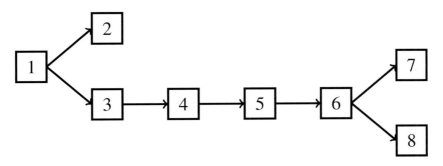

Fig. 1 Relationships among the sections

2   An Individual-Based Epidemic Model

Individual-based models,[4] or IBMs, are a good starting point for modeling because they are intuitive and can be implemented as an activity, thereby providing students with valuable direct experience [7]. Observing the effect of randomness inherent in IBMs helps students obtain a healthy skepticism that strengthens their ability to contextualize the results of deterministic models.

2.1   Model Description and Physical Simulation

Definition 2 An **individual-based model** consists of a database of *individuals*, each identified by one or more *attributes* that can change over time, and a set of *rules* that update the attributes of each individual at each time step.

Individual-based models can be implemented as a physical activity with actual people as the "individuals" in the model, a table-top simulation in which one modeler manages all the "individuals," or an automated simulation written with software such as MATLAB or NetLogo. A description of a physical activity for our disease model is available online [6]. See [5, 13] for more about individual-based models.

For our IBM, we assume a fixed population of N individuals, each of whom has a single attribute that indicates their current epidemiological status. There are four possible states:

1. "Healthy" individuals have not yet been infected;
2. "Presymptomatic" individuals have been infected and can transmit the disease, but do not show symptoms;
3. "Sick" individuals have been infected, can transmit the disease, and do show symptoms;
4. "Recovered" individuals are no longer sick, cannot transmit the disease, and cannot be reinfected.

We assume individuals move linearly through these states, although that assumption could be modified as a variation of the original model. We will sometimes refer to these states using the letters H, P, S, and R.[5]

The state attribute in the model can be identified in various ways. When enacting the model as a physical activity, the participants carry a set of four colored status cards—green for healthy, yellow for presymptomatic, red for sick, and blue for

---

[4] The term *agent-based model* is also common.

[5] These letters are chosen to help students build intuition before we introduce formal models. In standard epidemiology models, the symbol S is used to represent susceptible individuals, but in the current context, it makes more sense to use S for sick individuals.

recovered—and place the appropriate card on the top of their stack. In a computer implementation, the status can be a number: 0 for healthy, 1 for presymptomatic, and so on. The rules that govern the changes in states have been carefully balanced; they work best if there are at least 16 individuals in the population, with a starting point of two presymptomatic individuals in a small initial population or 2% in a large one, and the rest initially healthy.

Once the initial states have been assigned to the individuals, the simulation consists of consecutive time steps, each divided into phases:

1. Individuals in the population are randomly assigned to pairs. This can be done using a fully random approach, such as dealing out a deck of cards (each representing a member of the population) in pairs, running a web-based app that creates pairs from a list, or using a partially random "speed dating" structure that is more convenient for physical enactments because it takes less time [6]. In a fully automated implementation, only the healthy individuals need to be assigned a partner.
2. Healthy individuals who are paired with a presymptomatic or sick individual become infected with probability 5/6; in a physical enactment, this is easily done by rolling a die and equating a die roll of one with avoiding the infection. Those who do become infected advance from healthy to presymptomatic.
3. After all pairs have been checked for disease spread, any individuals who were presymptomatic at the beginning of the time step become sick, while those who were sick at the beginning of the time step become recovered.
4. The number of individuals in each status category is recorded for subsequent analysis.

The simulation proceeds from one day to the next, each day starting with new partner assignments and continuing with status updates. Eventually the simulation ends when all remaining individuals are either healthy or recovered. Once there are no presymptomatic or sick individuals remaining, there can be no further status changes.

This simple individual-based model captures many of the features of real disease spread. Individuals interact randomly and possible transmission encounters may or may not result in actual transmission. Other features are not so realistic. In reality, the amount of time spent in a presymptomatic or sick state can vary from one person to another. Most diseases have an incubation period of noticeable duration, necessitating a latent stage between the healthy and presymptomatic stages. In real disease settings, individuals have large numbers of daily contacts, each with a low probability of transmission. Our individual-based model obtains realistic results with a simple mechanic by reversing this pattern, with only one daily contact and a high probability of transmission.

The simplicity of this individual-based model also highlights the critical role assumptions play in modeling a phenomenon. In this context, the need for and outcomes of changing the model's assumptions can easily be understood, demonstrating the importance of consulting with experts to ensure that a model incorporates critical features and exhibits necessary behaviors. (Some of these assumptions will be considered in the exercises.)

2.2   Computer Simulation

While a physical simulation is a great way to build intuition, it is not a great way to study general model behavior. Instead, it is helpful to write a computer simulation for our individual-based model. We have provided two resources for this purpose.

Challenge Problem 1 provides detailed directions for creating a spreadsheet implementation of the HPSR individual-based model. This can be used (with modification) for the research projects in this section.

Spreadsheets have the advantage of familiarity and an intuitive structure. They are not generally used for serious modeling work because of two significant drawbacks: (1) the formulas that encode the model can only be viewed singly rather than as a package, and (2) they lack some essential programming functionality, such as the capacity to easily repeat computations with minor variations.

As an alternative to spreadsheet programming, we recommend that students use computer programs written for the MATLAB programming environment. The MAT-LAB environment has its own language, not much different from the better-known Visual Basic language, along with a large number of powerful, built-in functions that do complicated computations. Since few students who take mathematics courses at any level have any programming background, we provide MATLAB programs we wrote specifically to be accessible to beginners.

The Appendix contains a suite of three MATLAB programs that automate the model computations and obtain results from the output data. We present here a brief description of these programs. Most academic institutions have agreements that allow students to obtain MATLAB at little or no cost. For those that do not, there is free online software called Octave that is able to run most MATLAB programs [12].[6] We assume that students have acquired basic operational knowledge of how to load programs into the edit windows and how to run the programs from those windows.

MATLAB programs are of two types: scripts and functions. Scripts are self-contained lists of instructions written in syntax that can be understood by an interpreter that translates the instructions into computer code. These can be run from the MATLAB editor or an Octave window by clicking the *Run* menu item.[7]

Functions are special programs that automate some of the more complicated program tasks. These have a specific name along with a list of arguments and output variables. Part of the strength of MATLAB as an environment for scientific

---

[6] As of this writing, the programs all run successfully in Octave.

[7] In MATLAB, any changes entered in a script will be saved before the script is run. In Octave, you must manually save changes before rerunning a script.

computation is its extensive library of built-in functions, while another part of its strength is its facility for user-defined functions.

Functions cannot be run as separate programs, but are instead called from a command window or run from a script, just as one might use *y=cos(pi)*. The program suite for the IBM is based on a function *[H,P,S,R]=hpsr(b,N,P0,S0)*,[8] which automates the entire individual-based model computation and returns the time series of class counts. Function *hpsr* requires input values for the transmission probability (*b*), the total population (*N*), and the initial numbers of presymptomatic (*P0*) and sick (*S0*). It needs to be modified for Programming Level 2 projects to match changes in the model rules, but does not need to be changed for Level 1 projects.

The IBM program suite includes two scripts that make use of function *hpsr*.

- *HPSR_sim.m* uses *hpsr* to run one simulation and produce a plot of the results. The user needs to modify the *SCENARIO DATA* section for each run. The function call in the *COMPUTATION* section and some of the *OUTPUT* instructions will need to be changed for Level 2 projects.
- *HPSR_avg.m* collects data from multiple simulations and computes the mean and standard deviation of two key results: the maximum infected fraction and the final healthy fraction. The user needs to modify the *SCENARIO DATA* section for each run. The function call in the *COMPUTATION* section will need to be changed for Level 2 projects.

2.3   Section 2 Exercises

Exercise 1 (Calculus Level 0, Programming Level 0) Run the individual-based model three times using either a physical simulation, with people playing the roles of the individuals, or a table-top simulation. In each case, plot graphs of the daily class counts. Also record the number and identities of individuals who remained healthy throughout the simulation. Are those who do not get sick different in any meaningful way from those who do? In a real situation where an illness is spreading among a population, might there be a difference between those individuals who do not get sick and those who do? Discuss.

Exercise 2 (Calculus Level 0, Programming Level 0) The individual-based model uses four different states, each marked by a different color in the physical simulation, to describe the illness progression.

(a) Using the rules as described, explain which, if any, of the color distinctions are needed and which are not.

---

[8] By convention, we choose to give functions names that start with a lower-case letter, while script names start with a capital letter.

(b) Now suppose the rules for the physical simulation are changed so that half of the individuals who are sick choose to isolate. Explain which, if any, of the color distinctions are needed and which are not.

(c) How might the rules change if some of the individuals are vaccinated? How might this impact the meaning or number of the colored status cards?

(d) How might the rules change if individuals could contract the same illness more than once? How might this impact the meaning or number of the colored status cards?

Exercise 3 (Calculus Level 0, Programming Level 0) The individual-based model described in this unit assumes the duration of each phase of the illness is a single day. In reality, most illnesses do not progress this quickly. Visit the website [3] for information about the H1N1 virus, often referred to as "swine flu." Then develop a set of status cards and rules for an individual-based model of H1N1. Assume that the model would incorporate isolation of sick people. Would you still continue to use the same four states or would you need more? How many status cards would you use for each state? How many would you need for the full simulation? Would you need to incorporate additional die rolls into the model? Fully justify any assumptions made about the states of the illness and the duration of each state.

Exercise 4 (Calculus Level 0, Programming Level 1) Run the individual-based model program *HPSR_onesim* three times using parameter values that match your physical simulation in Exercise 1. Compare the results. Do they give convincing evidence that the computer simulation and physical simulation actually implement the same model?

Exercise 5 (Calculus Level 0, Programming Level 1) Run the individual-based model program *HPSR_avg* using parameter values that match your physical simulation in Exercise 1. Compare the results. Is this a good way to check that the computer simulation and physical simulation are actually implementing the same model?

2.4   Section 2 Challenge Problem

Challenge Problem 1 (Calculus Level 0, Programming Level 0) Create a spreadsheet implementation of a slightly modified version of the individual-based model described in the section. Here are some recommended guidelines:

1. Use row 1 to list the four parameters by symbol: the transmission probability b, the population size N, and the initial counts of presymptomatic P_0 and sick S_0. The values of these parameters go in row 2, beneath the parameter names. For a

first run, we recommend a transmission probability of 5/6 or 1 and a population of 1000 that includes 20 presymptomatic and 0 sick.

2. Use row 4 to enter headings for the data columns. Columns for the day and the four population classes (H, P, S, R) are a minimal requirement.

3. We recommend having three additional columns for the total population, the infectious population $I = P + S$, and the number of new infections.

   (a) The values in the total population column should be calculated using a formula that sums the four classes. This serves as a check on the spreadsheet formulas.

   (b) The values in the infectious column are just the sums of the P and S columns. The main benefit of this column is that a graph of the daily infectious count is very useful.

   (c) The new infections column is helpful because it is good to do calculations in several small formulas rather than one messy one. Without identifying individual pairings, we can estimate the fraction of healthy people whose "partner" in the simulation is infectious as the fraction of the total population that is infectious. Then we can assume that a fraction b of these healthy people with infectious partners become presymptomatic in the next time step.

4. Each of the columns needs to be given formulas to determine the values using the data in row 2. We recommend running the simulation for 30 days. The initial counts for classes H, P, S, and R can be calculated from the values entered in the cells for N, P_0, and S_0. Subsequent values in these columns can be calculated from the previous values along with the number of new infections in the last column.

5. Spreadsheets do computer calculations, which means that they produce numbers with many decimal digits. This makes it very hard to read the results. We recommend that you format all of the cells in the data table as "Number" with 0 decimal digits.

6. Raw numbers are hard to interpret. Your spreadsheet will be more useful if you make a graph. Use the scatter plot with only points because the data is discrete. Plot three data series: H, I, and R. Make sure your graph has axis labels and a legend. You can use the default color and style scheme; alternatively, you can match the color scheme of the card system by using dark green for Healthy, red for Infectious, and blue for Recovered. The graph will be easier to read if you use different styles for the points in each data series. One possibility is shown in the MATLAB program *HPSR_onesim.m*.

When you are finished, try using parameter values that match your physical simulation so that you can make sure the results are consistent. Alternatively, use a comparison with *HPSR_avg* with the recommended population size of 1000.

2.5   Section 2 Projects

In each of these projects, it is best to start by using the physical simulation to make sure that your model changes make sense. Then use the spreadsheet from Challenge Problem 1 and/or the MATLAB program suite to collect data.

Research Project 1 (Calculus Level 0, Programming Level 0 or 1) Study the effect of the transmission probability b on the results of the disease in the individual-based model. Hint: You can collect useful data for any value of b using the spreadsheet implementation of Challenge Problem 1 or by running *HPSR_avg*.

Research Project 2 (Calculus Level 0, Programming Level 0 or 2) Suppose patients with the disease are sick for two days instead of one. Modify the model to account for that possibility. Try keeping the transmission probability the same and also adjusting it downward by some amount. Determine what transmission probability yields results most like those of the original model with the larger transmission probability and one day of sick time. Hint: You can break the sick class up into two subgroups.

Research Project 3 (Calculus Level 0, Programming Level 0 or 2) As an extension of Research Project 2, suppose a fraction p of patients with the disease are sick for two days, while the remainder are sick for just one day. Design experiments with this model and describe and explain the results.

Research Project 4 (Calculus Level 0, Programming Level 0 or 2) Suppose individuals have two contacts per day instead of one. Modify the model and compare the results with the original model.

Research Project 5 (Calculus Level 0, Programming Level 0 or 2) Add isolation of the sick to the original model or one of the variations described in other projects. Note that you will need a parameter to represent the probability that an individual isolates when sick. The effect of isolation will depend on the value of this parameter.

3   Continuous-Time Dynamical Systems

While the individual-based model of Sect. 2 is discrete, most epidemiology models take the form of continuous-time dynamical systems. These systems consist of differential equations, which means that they are based on calculus. However, some of the background needed to understand these systems is not part of the standard calculus curriculum; conversely, very little of what does appear in a calculus curriculum is actually needed for our purposes. The goal of this section is to provide a self-contained introduction to dynamical systems that can be understood by students with a strong background in algebra and pre-calculus.

3.1   The Derivative

Standard first courses in calculus focus on the *derivative*. The text material includes the motivation for the derivative, its definition, rules for computing derivatives, and a variety of applications of the derivative. While all of this material is important in calculus, all we really need for dynamical systems is the geometric concept of the derivative. Here is a concise summary.[9]

Figure 2 illustrates a pair of functions, $x(t)$ and $y(t)$, that might represent the population size of a growing community. The function x shows *linear* growth, as would happen when individuals join the community at a fixed rate from the outside. That rate is the change in x divided by the change in t, which is the slope of the line in the graph. We can calculate the slope using algebra by picking two points, calculating the differences in their x and t values, and taking the ratio; here,

$$\text{slope} = \frac{\Delta x}{\Delta t} = \frac{0.8 - 0.6}{1.5 - 1} = 0.4. \tag{2}$$

The function y shows what is called *logistic* growth, which means that the growth is limited by the availability of resources. In this instance, the population grows by virtue of having more births than deaths, with no migration. It grows slowly at first

---

[9] See [9] for a more complete presentation.

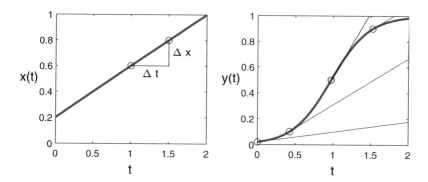

Fig. 2 A linear function $x(t)$ and a nonlinear function $y(t)$ (the thick curves), showing the slope and instantaneous slope (the thin lines) of the two functions at various values of t

because there are few potential parents, then increases more rapidly, and then levels off as the availability of resources or space becomes critical.

It still makes sense to talk about the slope of the graph of y, which we can think of as the slope of the line tangent to the graph at a particular point. A tangent line has only one known point, so we cannot calculate its slope with standard algebra; techniques for determining the slope of these tangent lines are developed in calculus. For our purposes, we can assign a symbol to represent the slope at a point t and focus on its interpretation. The symbol we use is dy/dt. which is similar to the notation $\Delta y/\Delta t$ for the slope of a line. The key difference is that we have to interpret dy/dt as a single symbol, not a quotient of two different things.

Definition 3 For a function $y(t)$ that has a smooth graph, the **derivative** dy/dt at any given point is the slope of the tangent line at that point, which represents the instantaneous rate of change of the quantity y.

3.2 Dynamical Systems

In a calculus text, the derivative of a function $y(t)$ is nearly always given as a function of the independent variable t. In mathematical modeling, we often have systems in which the rates of change are dependent only on the state of the system, which means that the derivative is given as a function of the dependent variables.

Definition 4 An **autonomous dynamical system** is a system of quantities whose rates of change in time are given as functions of the quantities themselves.

As an example, the dynamical system model

$$\frac{dy}{dt} = f(y) = 4y(1-y), \qquad y(0) = y_0, \qquad 0 < y_0 < 1, \qquad (3)$$

where y is a function of t, represents the spread of misinformation (which can be considered as similar to the spread of a disease). Everyone is either "susceptible" to the misinformation or "infectious" with it, and there is no "recovery." The dependent variable y is the fraction of the population that is infectious. This fraction increases for any current fraction y between 0 and 1 according to the function $f(y) = 4y(1 - y)$.

How should we try to understand the dynamical system (3)? The simplest way is to use the information shown by a graph of the known function $f(y) = 4y(1-y)$ for the range of values from $y = 0$ to $y = 1$, not to find a formula for the function $y(t)$, but to check slopes on a sketch of the graph of y vs t. If we know the current value of y at any point in time, then we can use the graph of f to identify dy/dt, which is the slope of the graph of $y(t)$ at that particular time.

Figure 3 shows two graphs, one of the function $f(y)$ and one of the function $y(t)$. The first graph gives us information we can use to verify the second graph. Suppose the value of y is initially 0.01. From the left panel of Fig. 3, we can see that the value of f, and hence of dy/dt, is small and positive. So y is increasing, but with a fairly flat rise. The first marked point in the right panel shows the same value of y and a tangent line that has the slope found from the f axis in the left panel. As y increases, the graph of f tells us that y will continue to rise with increasing slope. In other words, it will curve upward as it increases. The second marked point is at $y = 0.1$ and shows the value of dy/dt to be a little less than 0.4. The corresponding point in the panel on the right has $y = 0.1$ and shows the tangent line with the correct slope. This behavior starts to change when we reach $y = 0.5$, since that is where f achieves its maximum value. This point corresponds to the steepest slope on the graph of $y(t)$. After that, the function f is still positive, but its value decreases as y increases further. The graph of $y(t)$ continues to increase, but with a flattening, or downward curvature. The fourth marked point has the same value of f as the second, which means that the slopes of the graph of y at those two points are the

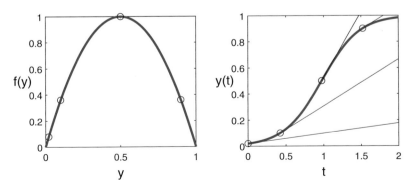

Fig. 3 The function $f(y) = 4y(1-y)$ and the solution of $dy/dt = f(y)$ with $y(0) = 0.01$

same. Note that the graph of $y(t)$ can never cross the threshold value $y = 1$ because f is 0 when $y = 1$, meaning that y is no longer increasing.

The graph of y in the right panel was created by applying calculus to obtain the solution formula for the system (3); however, it could instead have been obtained from a numerical simulation using a method presented in Sect. 8. Without one of these methods, we would not have been able to get the times right for the four marked points; however, we would still have been able to get the right slopes for each value of y, which give us the correct overall shape for the graph.

To summarize: a differential equation of the form $dy/dt = f(y)$ specifies the slope of the graph of y as a function of the value of y. This information, along with an initial condition, is sufficient to define the graph, which we can roughly sketch with nothing more than logical statements about how f changes as y changes.

It is not a large leap to appreciate that this same idea will work for a system of four differential equations for variables S, E, I, and R, where the derivatives are known functions of the four variables. The rates of change for each variable will be functions of more than one state variable, so we will not have a graph like the one in the left panel of Fig. 3; however, we will have the formulas that calculate the slopes from the state of the system.

3.3 Section 3 Exercises

Exercise 6 (Calculus Level 1, Programming Level 0) The differential equation $dy/dt = ky$, where k is any positive constant, models exponential growth. In contrast, explain why Eq. (3) and the graph of its solution (see the graph on the right in Fig. 3) is an appropriate model for a scenario in which there is a constant but limited amount of available resources.

Exercise 7 (Calculus Level 1, Programming Level 0) As noted in Exercise (6), the differential equation in (3) is used to model the growth of a population with a fixed resource limitation. In models representing the interaction between predator and prey populations, the function $y(t)$ can represent the predator while the prey can be considered the limited resource. For what situations would the equation in (3) be appropriate for modeling the population of a predator species for a given prey? For what predator-prey situations would it *not* be appropriate? Justify your claims.

Exercise 8 (Calculus Level 1, Programming Level 0) The dynamical system model

$$\frac{dy}{dt} = f(y) = 4y(1 - y) - 3y, \quad y(0) = 0.01$$

represents the infectious population fraction for a disease in which recovered individuals can be reinfected (like the common cold). Plot the graph of f and use it to sketch a possible graph for $y(t)$. (Keep in mind that y represents a population fraction.) Check your graph by plotting the function

$$y(t) = \left(4 + 96e^{-t}\right)^{-1},$$

which is the solution of the differential equation problem.

Exercise 9 (Calculus Level 1, Programming Level 0) Repeat Exercise 8, but with $y(0) = 0.5$. [Keep in mind that you will not be able to use the given solution formula as a check.]

Exercise 10 (Calculus Level 2, Programming Level 0) Show that the function

$$y(t) = \left(4 + 96e^{-t}\right)^{-1}$$

solves the initial value problem

$$\frac{dy}{dt} = f(y) = 4y(1 - y) - 3y = y - 4y^2, \quad y(0) = 0.01.$$

Hint: If we define $u(t) = 4 + 96e^{-t} = y^{-1}$, the right side of the differential equation becomes $(uy)y - 4y^2 = (u - 4)y^2$.

Exercise 11 (Calculus Level 2, Programming Level 0) Show that the function

$$y(t) = \left[\frac{b}{a} + \left(\frac{1}{y_0} - \frac{b}{a}\right)e^{-at}\right]^{-1}$$

solves the initial value problem

$$\frac{dy}{dt} = ay - by^2, \quad y(0) = y_0.$$

Hint: See Exercise 10.

4 Dynamical System Models

Model development is as much an art as it is a science. The science is in the employment of logical structures to serve as the model framework, while the art is in finding creative ways to conceptualize the motivating real-world scenario.

In Sect. 2 we saw one class of models—individual-based models. They have the advantage of easy conceptualization and inclusion of randomness, but they have some drawbacks as well:

1. They have to be run many times to get a picture of the expected behavior.
2. They become very complicated when more realistic features are added, such as random durations of incubation and infectivity.
3. They do not lend themselves to mathematical analysis.

Most mathematical modeling in epidemiology instead uses dynamical system models, which we explore in the remainder of the chapter. Dynamical system models in epidemiology have a long history [2]. The class of models most commonly used today was first introduced in a seminal paper by W.O. Kermack and A.G. McKendrick published in 1927 [8].

Dynamical system models are based on the same principle as accounting. If you want to know how much money you have in different budget categories, you can (1) keep the money in different boxes and count it every day, or (2) keep track of your income, expenditures, and internal transfers, and use these amounts to calculate daily totals. In epidemiology, the budget categories are replaced by variables that indicate the current totals of people in various epidemiological states, such as "Susceptible" (which we called "healthy" in the individual-based model) or "Infectious" (which often includes both presymptomatic and sick individuals).

Mathematical epidemiology is a rich subject for which one can find books at levels from advanced undergraduate to research level.[10]

4.1 Classification of Dynamical System Models

Several features are needed to classify epidemiology models. We present these in rough order of importance.

1. Disease Type
   Infectious diseases can be divided into two subgroups, based on the transmission mechanism.

---

[10] Space does not permit an exhaustive list of references. For lower-division undergraduates, we recommend the oddly titled book by Smith [15] because it builds up to a sophisticated level from a very humble beginning. For advanced undergraduates, the paper by Blackwood and Childs [1] is a good place to start.

- **Person-to-person** transmission involves either direct transmission through physical contact or indirect transmission through the environment, such as droplets that enter the air through sneezing or coughing.
- **Vector-borne** transmission is required for diseases in which the pathogen has a complicated life cycle that requires multiple host species. An example is malaria, which is caused by a protozoan that lives part of its life in humans and part in mosquitoes.

Most infectious diseases which affect humans fit neatly into one of these two categories, but some are less clear. The plague that swept through much of the world in the medieval period had two forms: bubonic, which humans contracted from fleas, and pneumonic, which was passed from person to person via droplets in the air.

In this chapter, we consider only diseases with person-to-person transmission.

2. Time Frame

- **Epidemic** models are designed for the short term. Their distinguishing characteristic is that there are no mechanisms for replenishment of susceptible people. Like a forest fire that must eventually run out of fuel, an epidemic scenario necessarily ends with no remaining disease. Study of epidemic models focuses on rate of spread, change in the number of infected individuals over time, and final susceptible population size. Analytical methods are available for some results in the simplest cases, but usually simulations are required. Epidemic models can be either deterministic (no randomness) or stochastic (having some random elements), depending on the level of detail desired for quantifying the various disease processes.
- **Endemic** models are designed for the long term. They always include at least one mechanism for replenishment of susceptibles, typically birth of susceptible individuals, but there can also be processes such as immigration or gradual loss of immunity. There are also diseases, such as the common cold, that do not confer subsequent immunity at all, perhaps because there is a plethora of strains or very rapid mutation. The long-term behavior of endemic models can be obtained using computational and analytical methods.

Many treatments of mathematical models in epidemiology focus primarily on endemic models, because these are the ones for which mathematical analysis can make the strongest conclusions. In this chapter, we instead focus on epidemic models to reflect their importance in the COVID-19 dominated world of 2020 and beyond.

3. Population Constancy

The total population can be fixed by omitting demographic processes altogether, by making sure birth and death rates are equal with no additional mortality or migration, or, in the case of epidemic models, counting deceased individuals as equivalent to recovered ones. This simplifies the model as compared to the more common case of variable size populations and is typically done for all epidemic models as well as some endemic models.

4. Classes

Although the choice of classes is no more fundamental than the choice of time frame, it is traditional to name models according to the list of classes used in them. Each class tracks the number of individuals with that particular disease status. Our starting point is the SEIR model, where

- S is *Susceptible*, for individuals who are at risk of catching the disease. (We called this group "Healthy" in the individual-based model.)
- E is *Exposed*, for individuals who have been infected but are not yet infectious. This is a poor choice of term–"latent" is more accurate. In everyday language, we would say that a person has been "exposed" when they have contacted an infectious person, regardless of whether they have caught the infection, but in epidemiology all members of the "Exposed" class have been infected. As a compromise between accuracy and consistency, we will retain the class symbol E but use the term *Latent* as the class name. (The individual-based model did not have a latent class.)
- I is *Infectious*, for individuals who can transmit the disease. In the basic SEIR model, infectious individuals may or may not have symptoms. (In the individual-based model, the infectious class was divided into presymptomatic and symptomatic subgroups.)
- R is *Removed*, for individuals who are not currently infectious and are immune from further infection. This can include individuals who have recovered and individuals who are still sick but no longer infectious. Including deceased individuals as "removed" is convenient if death by the disease is the only demographic process, as then the population can still be thought of as "constant" in the sense that the class system is then closed.

Many disease models include additional or different classes based on features of the particular disease being studied. There could be isolated infectives[11] and quarantined susceptibles, and infectives could be further subdivided into asymptomatic and symptomatic. A COVID-19 model should distinguish between asymptomatic and symptomatic infectives and between unconfirmed, confirmed, and hospitalized cases.

5. Processes

In addition to a list of epidemiological classes, a model needs a list of processes that cause individuals to move from one class to another. In an SEIR model, there must be a transmission process that moves susceptible individuals into the latent class, an incubation process that moves latent individuals into the removed class, and a removal process that moves infectious individuals into the removed

---

[11] Technically speaking, the word "infective" does not appear in official dictionaries; however, it is a much better term than the alternative "infectious individual."

class. There could be other processes, such as vaccination, isolation, and loss of immunity. Some of these are considered in the projects.

6. Dynamical System Type

Discrete-time models are based on algebra, while continuous-time models are based on calculus. Discrete-time models can seem more intuitive, but continuous dynamical systems have much better mathematical properties. We will follow the more common practice of using continuous models.

5   Building the SEIR Epidemic Model

Figure 4 shows a schematic of the structure of the standard SEIR model. The four population classes are related by three processes:

1. A **transmission** process moves susceptible individuals into the latent class (E).
2. An **incubation** process moves latent individuals into the infectious class.
3. A **removal** process moves infectious individuals into the removed class.

Notice what is missing:

- We have not explicitly included disease-induced mortality, but that does not mean we have not allowed for it. It is customary in epidemic models to make no distinction between people who have died from the disease and people who have recovered. This undoubtedly seems horrible to the novice modeler, but it is actually very instructive. From a *human* perspective, we want to distinguish between healthy people, sick people, recovered people, people with ongoing deleterious effects, and people who died. But our goal is to make an *epidemiological* model. There is no epidemiological distinction between infectious people who are sick and those who are not, or between removed people who are healthy, still suffering illness, or deceased. These human interest features can be added to the base epidemiological model during analysis.

- We have no birth or natural death processes and no migration. Obviously these demographic processes do occur during a disease outbreak. However, on an epidemic time scale of weeks or months, births and natural deaths only change the class counts to a limited extent. Remember that our model is not intended to exactly match reality. We will see assumptions in the quantification of the epidemiological processes that will yield more quantitative error. Some of these will be explored as research projects.

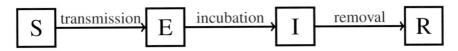

Fig. 4 The SEIR epidemic model in words

- We have not included any modifications for public health measures, such as vaccination, isolation of infectious individuals, or quarantine of individuals through contact tracing. The end-of-chapter projects will focus largely on adding possible modifications. There are many ways to do this, not all of which have been thoroughly explored. This is part of the art of mathematical modeling, and it offers students an opportunity to do original research by being the first to think of a new idea.

From the schematic diagram, we can immediately write a set of conceptual equations for the changes in the class sizes. The susceptible population is affected by only one process–the transmission process, which decreases the population:

$$\frac{dS}{dt} = - \text{ transmission rate}. \tag{4}$$

Similar equations describe the overall rates of change for the other classes:

$$\frac{dE}{dt} = \text{ transmission rate} - \text{ incubation rate}, \tag{5}$$

$$\frac{dI}{dt} = \text{ incubation rate} - \text{ removal rate}, \tag{6}$$

$$\frac{dR}{dt} = \text{ removal rate}. \tag{7}$$

Now suppose we define N to be the total population; that is,

$$N = S + E + I + R. \tag{8}$$

The change in N is the sum of the changes in the four components. Since each process increases one class at the expense of another, the overall population stays the same (given the interpretation of "deceased" as being epidemiologically equivalent to "recovered"). This means that we could omit one of the change equations; for example, we could omit the R equation and instead use the N equation to calculate R. This is commonly done when focusing on the mathematical analysis of a model. However, when doing a computer simulation, it is good practice to use the change equations to calculate all class counts and then use the N equation as a check for possible error in the program.

While we could use N to represent the actual population, it is usually best to take $N = 1$, which means that the class sizes are used to measure fractions of the total population rather than raw numbers. Fractions of a total are meaningful with no additional context, whereas class counts are only meaningful in comparison to the total population.

5.1   Quantifying the Processes

To complete the SEIR model, we need to find mathematical formulas for each of the three processes in the compartment diagram and conceptual equations. There are standard choices that are used in nearly all circumstances. These standard choices are based on some questionable assumptions, so it is important to understand the formulas in some detail. (See [15] for a similar discussion with much more detail.)

The first step in quantifying each of the processes is deciding what variables must be considered. Take removal as an example. If a lot of people are infectious now, then surely people will leave the infectious class at a high rate; however, if nobody is infectious, then nobody can be removed from the infectious class and the removal rate is 0. Thus it is reasonable to assume the removal rate depends on the size of the infectious class. What about the other three classes? Does the rate of removal from the infectious class depend on how many people are susceptible, how many are latent, or how many are already removed? The answer is "no" in all cases. Whether an infectious person is removed in a given day depends only on what is happening inside that infectious person, not on the states or changes in state of anyone else. The removal process is an example of a *transition*: a process that happens to each individual person in the departing class, independent of anyone else. Similarly, the incubation process is a transition. Each latent individual becomes infectious based on their own internal changes, independent of what is happening to others. The more people in the class that undergoes the transition, the greater the rate at which the transition occurs at the population level.

The transmission process is fundamentally different from the transition processes. Susceptibles are necessary for transmission to occur. But unlike incubation, which happens to latent people without any interaction, transmission does not happen independently. You cannot get the disease without getting it from someone who is infectious.[12] Thus, the formula for the transmission process is going to need to use both the susceptible and infectious class counts. Accordingly, we consider transitions and transmission separately.

5.1.1   Transition Processes

Let us start with the incubation process. If we compare two moments in the history of an outbreak, and there are twice as many latent people in moment A than in moment B, it seems reasonable that the rate of incubation is twice as great at moment A. We, therefore, assume that the incubation rate is proportional to the latent class size.[13] This assumption tells us the mathematical form of the rate formula, but with

---

[12] Some diseases are spread through the environment, but the disease agent gets into the environment from infectious individuals. Epidemiologically, it does not matter if you get the disease by shaking hands with someone or by breathing in disease agents the other person left behind when leaving a room.

[13] This can be shown to be equivalent to the assumption that the incubation times are exponentially distributed, with the rate constant given by the reciprocal of the mean time.

a proportionality rate constant that is specific for each disease. It is customary to use lower-case Greek letters for this and other rate constants. The specific rate constant symbol for a particular process varies from one author to another. We will use the Greek letter "eta" for the incubation rate constant. Thus, we have

$$\text{incubation rate} = \eta E. \tag{9}$$

The removal process is analogous to the incubation process in that its rate is proportional to the size of the class it is leaving. Using the Greek letter "gamma," we have

$$\text{removal rate} = \gamma I \tag{10}$$

So far this may seem straightforward, but there is actually a lot more that needs to be said. As a thought experiment, suppose that the entire latent class is made up of people who just got infected yesterday. Assuming that the rate is proportional to the class size, the rate of incubation will be largest on day 1, when the entire cohort is still latent. This is a very suspicious claim. If it takes an average of 5 days, then most people will become infectious in 4, 5, or 6 days. It seems likely that nobody would become infectious in the first day, as that would require the disease to progress 5 times as fast as the average. The naive assumption that the incubation rate constant for latent patients will be the same each day is a significant conceptual error, far worse than the minor error of neglecting the small number of births that occur during the scenario. The reality is far more complicated than what we are assuming in our model.

In spite of the apparent seriousness of this conceptual error, there is some reason to think that it will not actually matter. Most of the time, the latent class will consist of people who were infected at different times. If our naive model predicts that someone who is newly infected suddenly becomes infectious today, while someone who has been infected for 6 days already does not, these errors should average out at the population level. This is really what we are assuming. While this issue is largely ignored in epidemiological modeling, we should keep it in mind whenever we are tempted to take our results too seriously. Exploration of this issue makes a nice research project (given here as Research Project 8, with background from Exercise 13).

5.1.2   The Transmission Process

Suppose each infectious person encounters a fraction c of a population of size N per day. Assuming that all possible encounters are equally likely, we can expect that a fraction S/N of these encounters are with susceptible individuals. This means that each infectious person has cS encounters with susceptible individuals. If each encounter has a probability p of resulting in a transmission, then we can expect one infectious person to transmit the disease to pcS individuals per day. If we have I infectious individuals rather than 1, the total rate of transmission is $pcSI$. We do not need individual factors p and c, so we can write the transmission rate as

$$\text{transmission rate} = \beta SI, \tag{11}$$

where β is the product of the contact fraction and the transmission probability. This formula can be used regardless of whether we are measuring the population classes as counts of people or fractions of the total population, but the value for β is different in the two cases. This is one of several reasons why it is hard to identify a value for this parameter. This issue will be addressed later.

As with the transition formulas, there are hidden assumptions in the simple formula for the transmission rate. If you reread the explanation, you should notice that it does not distinguish individuals from each other. In reality, we live in a social network where we have frequent contacts with some people and no contact with nearly everyone else. Some of us are at the center of our network and have lots of contacts, while others have far fewer contacts. In the COVID-19 pandemic, some people deliberately reduced the number and frequency of contacts, while others did not. Additionally, infectious individuals vary in their level of infectivity and susceptibles vary in their level of susceptibility. As with transition processes, we are assuming that we can use averages. On the other hand, it was easy to come up with a thought experiment in which the transition formula gives a value that is too high or too low, even averaged over the whole population at any given time. It is much harder to do that for the transmission process. While there may be some error introduced in an epidemiology model by the overly simple assumptions about transmission, it is almost certainly less error than is introduced by the assumptions about transitions.

5.2   The Final Model

Now that we have chosen formulas for the rates of the processes, it is helpful to redraw the compartment diagram, using those rates as the labels on the arrows (Fig. 5).

Working directly from the diagram gives us the differential equations that describe the rates of change in terms of the state of the system. Full model specification also requires initial conditions. We will use lower-case letters for these, as R_0 is often used for something entirely different from the initial condition parameters. The final model is

$$\frac{dS}{dt} = -\beta SI, \qquad S(0) = s_0 > 0; \tag{12}$$

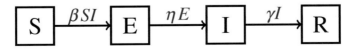

Fig. 5 The SEIR epidemic model in symbols

$$\frac{dE}{dt} = \beta SI - \eta E, \qquad E(0) = e_0 \geq 0; \tag{13}$$

$$\frac{dI}{dt} = \eta E - \gamma I, \qquad I(0) = i_0 \geq 0; \tag{14}$$

$$\frac{dR}{dt} = \gamma I, \qquad R(0) = r_0 \geq 0; \tag{15}$$

where

$$s_0 + e_0 + i_0 + r_0 = 1, \qquad e_0 + i_0 > 0. \tag{16}$$

These last two requirements ensure that the initial population total is 1 and that there are some infected people to get the outbreak started.

5.3   Section 5 Exercises

Exercise 12 (Calculus Level 1, Programming Level 0) Construct a compartment diagram similar to that of Fig. 5 for the class and process structure that matches the individual-based model of Sect. 2.

Exercise 13 (Calculus Level 0, Programming Level 0) Suppose there are two latent stages, E1 and E2, each with the same transition rate constant. Newly infected individuals are in class E1, pass to class E2 with a "partial incubation" process having rate constant σ, and then pass on to class I with a second partial incubation process also having rate constant σ. Prepare a compartment diagram like that of Fig. 5 for this SEEIR model.

Exercise 14 (Calculus Level 2, Programming Level 0) Prepare a compartment diagram like that of Fig. 5 for a model in which a fraction p of infectious individuals are asymptomatic (class A) and the remainder are symptomatic (class I). Assume that asymptomatic individuals do not subsequently become symptomatic. In completing the diagram, keep in mind that both classes in this SEAIR model are infectious.

6   Modeling

In addition to variables and equations, mathematical models contain *parameters*. An understanding of parameters is critically important in mathematical modeling.

Definition 5 Parameters are quantities that function either as variables or constants, depending on the context.

The critical importance of parameters is illustrated in the schematic diagram of modeling shown in Fig. 6. In developing the model, we had time as the independent variable and dependent variables S, E, I, and R. In this *narrow* view, the parameters are constants. This is the realm of mathematics, be it analytical calculation (algebra and/or calculus) or numerical simulation. We can do analysis without parameter values, but we treat them the same way we treat the constant π; that is, as numbers denoted by symbols rather than numerals. Numerical simulation requires specific values for the parameters to produce numerical approximations for the rates of change at specific times.

Most important modeling work takes place in the *broad* view. Here is where we ask questions that we hope will be useful in interpreting the scenario that inspired the model rather than simply calculating results. In the broad view, we can study the effects of changes in the parameter values, asking questions such as "If two diseases have different incubation periods, how does that difference affect the outcomes of the model?" The outcomes can be whatever we are most interested in: for example, the maximum size of the infectious class, the day on which that occurs, and/or the total number of people who get the disease during the outbreak. Such information could be used to estimate the danger of running out of hospital space and the total number of deaths, for example.

The function concept is helpful in understanding how mathematical models work. In the narrow view, the class counts are functions of the number of days since the start of the outbreak. In the broad view, the outcomes are functions of the parameter values. Both the narrow view and broad view functions differ in an important way from the functions you are used to seeing in math classes. Functions in math classes are nearly always defined by explicit formulas, such as the formula

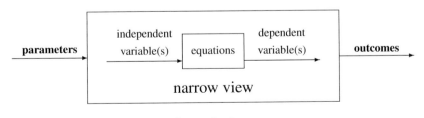

Fig. 6 Narrow and broad views of mathematical models

for the incubation rate (9). In mathematical modeling, functions are defined in much more subtle ways; for example, as the solution of a mathematical problem rather than as a mathematical formula. Conceptually, however, the function idea is the same regardless of whether the function is computed from a formula or as the result of a numerical procedure. It is best to think of functions in terms of their graphs rather than their formulas. Sections 7 and 8 develop tools for working with the functions in models. In this section, we focus on the specification of parameters.

6.1 Identifying Parameter Values

Initial condition parameters are determined by the specific scenario we have in mind. In practice, we will assume that the initial fractions of removed, latent, and infectious individuals are known, with everyone else susceptible. How we do this depends on the starting point for the scenario. For the initial introduction of a disease to a community, we assume that nobody is immune or infectious and that just a small fraction of people are latent. Other situations will be discussed later.

The three process parameters represent the quantitative characteristics of the particular disease being modeled. Often we have a specific disease in mind, so it is important to be able to use measurable properties to determine the values.

Dimensional analysis provides clues for the meaning of parameters. Things that can be compared in an equation or added have to have the same dimension, and the dimension of a product is the product of the dimensions. The terms dI/dt and γI must have the same dimension of people/time, so γ has dimension 1/time. What is not clear without mathematical development is how to know what time γ is the reciprocal of. The theorem given here is proven in Challenge Problem 2.

Theorem 1 *A transition rate parameter is the reciprocal of the average amount of time required for the transition process.*

While simple, this rule is actually very subtle. In order to derive it, we have to solve a calculus problem; hence, we defer this to the exercises.

Let t_L be the mean incubation period and t_I the mean infectious duration, which are generally known to a reasonable approximation. Then the transition rate parameter rule gives us

$$\eta = \frac{1}{t_L}, \qquad \gamma = \frac{1}{t_I}. \tag{17}$$

Note that the values of these parameters do not depend on whether we use population counts or population fractions for the state variables.

The transmission rate parameter β poses a special problem. It is not actually a fundamental disease property because its numerical value depends on how we define population size. It does not have the dimension of 1/time, so it cannot be the reciprocal of a measured time. Fortunately, there is another parameter that represents the infectiousness of a disease.

6.2   The Basic Reproduction Number

The *basic reproduction number*, which is given the symbol $\mathscr{R}_0$ and read as "R-nought," is the fundamental measure of the infectiousness of a disease.

Definition 6 The **basic reproduction number** is the average number of secondary infections brought about by one infectious person in a population that is wholly susceptible.

This definition of $\mathscr{R}_0$ sounds complicated, but it is much simpler if we break it down into parts. The transmission rate formula (11) tells us that the average number of secondary infections *per day* in a population of any composition is βSI. If that population is wholly susceptible, then we get an average of βNI secondary infections per day. That is the number produced by the whole infectious class; with $I = 1$, we see that "the average number of secondary infections *per day* brought about by one infectious person in a population that is wholly susceptible" is βN. To get the basic reproductive number, we just need to take into account that one infectious person has, on the average, t_I days in which to produce secondary infections. Total is rate times time, so the basic reproduction number is

$$\mathscr{R}_0 = \beta N t_I. \tag{18}$$

While we have calculated the basic reproduction number in terms of β, the actual use of this formula is usually to determine β from $\mathscr{R}_0$. Of course this means that we need to be able to determine $\mathscr{R}_0$ by some other means. You can look up values for well-known diseases. The issue of how to estimate $\mathscr{R}_0$ for a novel disease is critically important to accurate modeling results; this will be addressed in Sect. 7.

Note that the value of $\mathscr{R}_0$ is independent of the units used for population class sizes. If we change N, it is the value of β that makes a corresponding change. Given that we have chosen $N = 1$ by design, we can rearrange (18) and replace t_I with $1/\gamma$ to obtain

$$\beta = \gamma \mathscr{R}_0. \tag{19}$$

We can, therefore, specify a particular disease using t_L, t_I, and $\mathscr{R}_0$ as the three fundamental disease parameters and use (17) and (19) to calculate the parameters that appear in the model.

Before we move on, this is a good time to think about what the basic reproduction number tells us. Suppose the first generation consists of 10 infectious individuals. If $\mathscr{R}_0 = 3$, then each infectious person will generate an average of three new infections, for a total of 30 in the second generation. The epidemic will grow explosively as long as the population remains largely susceptible. Only when most of the population has been infected will we stop seeing more infections in the next generation than the previous one. In the end, nearly everyone will have got the disease.

Instead, if $\mathcal{R}_0 = 1.1$, there will be an average of 11 in the second generation. This is enough to keep the outbreak growing for a little while, but at a much slower rate than if $\mathcal{R}_0 = 3$, and a much smaller decrease in the susceptible population will be enough to stop the disease. Continuing with the comparison, if $\mathcal{R}_0 = 0.9$ then the second generation will average 9 individuals. We see that the disease is unable to get a foothold. Thus, $\mathcal{R}_0 = 1$ is a critical value—a disease can only cause an epidemic outbreak if $\mathcal{R}_0 > 1$.

We can now appreciate why COVID-19 is such a serious problem. The most common infectious diseases prior to COVID-19, the common cold and influenza, have $\mathcal{R}_0$ values on the order of 1.5–3. It is not uncommon for a person to have known exposures to someone with the flu and not get the disease. In contrast, the standard twentieth century childhood diseases of measles, chicken pox, and mumps have $\mathcal{R}_0$ values of 10 or more. Before the development of the vaccines for these diseases, virtually everyone who was exposed caught them.[14] The basic reproduction number for the original strain of COVID-19 has been reported to be in the range from 2.5 to 8. Our best estimate is $\mathcal{R}_0 = 5.7$ [14], which is a very large value.[15] When you run simulations, you will see that in a society that completely ignores the threat, almost the entire population will get the disease in less than 2 months.

6.3   Section 6 Exercises

Exercise 15 (Calculus Level 0, Programming Level 0) Estimates of the basic reproduction number and durations for incubation and infectiousness can be found online for many illnesses. Use published information about H1N1 flu [3] to estimate reasonable parameter values for η, γ, and β. Be sure to justify all assumptions and cite all references.

Exercise 16 (Calculus Level 0, Programming Level 0) Repeat Exercise (15) using the infectious disease of your choice. Justify all assumptions and cite all references.

---

[14] Curiously, neither of the chapter authors, both growing up before there were vaccines for these diseases, was ever diagnosed with chicken pox. Nevertheless, one of the authors is now known through detection of antibodies to have had the disease, and the other almost certainly had the disease as well because of repeated exposures.

[15] As of July 2021, the delta variant is just becoming dominant, with a basic reproductive number probably between 8 and 10. Given this value, virtually everyone who is not vaccinated will get the disease.

6.4 Section 6 Challenge Problem

Challenge Problem 2 (Calculus Level 2, Programming Level 0) Assume a population of newly infectious people and let $I(t)$ be the fraction of this cohort that is still infectious at time t, so that

$$\frac{dI}{dt} = -\gamma I, \quad I(0) = 1.$$

Use your knowledge of derivative formulas to identify the function $I(t)$ that satisfies these two requirements. Rewrite the result by solving it for t; this gives you the time at which the cohort has been reduced to a particular size. Now use calculus to determine the average value of t over the interval $0 \leq I \leq 1$. Explain why this proves Theorem 1.

7 Model Analysis

It is helpful to frame the model analysis by using the idea of a mathematical model as a function that determines outcomes in terms of the parameters. The details are shown in Fig. 7 as a schematic diagram.

There are six input parameters, the three that define the disease properties and three that define the starting point of the scenario. There are a number of possible outcomes we could be interested in. We focus on five of these:

1. The early stage exponential growth coefficient, λ (this will be defined shortly), which tells us how rapidly the epidemic grows at the beginning;
2. The maximum infectious class size, I_{max}, which tells us how much impact the epidemic will have at the worst point;
3. The time at which the maximum infectious class size occurs, t_{max}, which tells us how much time there is to prepare for the peak;
4. The ending susceptible population, denoted as s_∞, which tells us how much of the population does not contract the illness and will be at risk in a subsequent scenario;

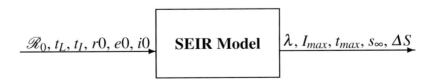

$$\mathcal{R}_0, t_L, t_I, r0, e0, i0 \longrightarrow \boxed{\textbf{SEIR Model}} \longrightarrow \lambda, I_{max}, t_{max}, s_\infty, \Delta S$$

Fig. 7 Schematic diagram of the SEIR model as a function

5. The total population infected during the scenario, $\Delta S = s_0 - s_\infty$. If we know what fraction of infected people die, we can use ΔS to calculate the total number of deaths in a given initial population.

Our outcomes will need to be determined by a variety of methods. We can use analytical methods (exact calculations using algebra or calculus) to compute λ and to derive an algebraic equation for s_∞. That equation cannot be solved using analytical methods, but it can be solved with numerical methods (approximate calculations using a computer program). Most epidemic model outcomes, such as I_{max} and t_{max} in the SEIR model, can only be determined by a fully numerical method. In this section, we consider λ and s_∞; the following section discusses numerical simulation of the full model.

7.1   Early Phase Exponential Growth

Analytical methods and numerical methods are complementary in many ways. One of these is that the behavior illustrated by numerical simulations can yield conjectures that can subsequently be confirmed by analysis. Figure 8 shows the results of a simulation using an initial condition of no infectious or removed individuals and only one latent individual per 10K population. The plot on the left shows the typical pattern of an epidemic outbreak. It takes a while to get started, but then the infection grows rapidly. Both the latent and infectious classes reach a peak and then drop off to 0, with the latent peak occurring earlier in time than the infectious peak. The plot on the right shows some very important detail that can only be seen on a logarithm plot. There is a very fast initial phase during which the latent population is decreasing. This is because we started without any infectious individuals, so new transmissions had to wait until the first batch of latent individuals became infectious. Then there is a significant period of time during

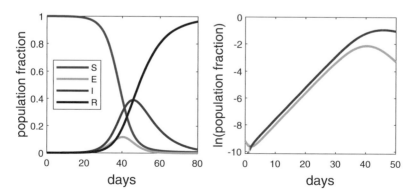

Fig. 8 Simulation results for the continuous model with $R_0 = 5$, $t_L = 2$, $t_I = 10$, $e_0 = 0.0001$, $i_0 = r_0 = 0$

which the graphs of $\ln(E)$ and $\ln(I)$ are linear *and parallel*. Only as the latent population comes close to its peak do the graphs of $\ln(E)$ and $\ln(I)$ begin to curve downward. These same features appear with any realistic choices for the disease parameters, as long as the initial fractions of the exposed and infectious classes are small. From this graph, we can make the following conjecture:

- After a short initial adjustment phase, there is a period in which the logarithms of the infected classes are linear with a common slope λ.

Mathematical exploration of this insight leads to the following result (see Challenge Problem 3):

Theorem 2 *Suppose the initial populations of the infected classes are small compared to that of the susceptible class. Then the SEIR model shows an extended exponential growth phase with*

$$I \approx I_0 e^{\lambda t}, \quad E \approx \rho I_0 e^{\lambda t}, \quad S \approx s_0 \approx 1 - r_0, \tag{20}$$

where λ is the positive solution of the equation

$$(\lambda + \eta)(\lambda + \gamma) = \eta \gamma s_0 \mathcal{R}_0, \tag{21}$$

and

$$\rho = \frac{\lambda + \gamma}{\eta}. \tag{22}$$

Theorem 2 has two very important consequences. First, it gives us a way to estimate $\mathcal{R}_0$ from early data on the infectious class population. From data for $\ln(I)$, we can estimate the value of λ and then use (21) to estimate $\mathcal{R}_0$. This is the best way to estimate $\mathcal{R}_0$ for a novel disease, like COVID-19. In late March of 2020, the "accepted" value of $\mathcal{R}_0$ for COVID-19, obtained by statistical analysis of known transmissions, was 2.6. One of us (GL) used (21) to obtain an estimate of 5.0 at that time. Subsequently, our current best guess of 5.7, obtained using a method similar to ours, was published in July 2020 [14].

The other important consequence of Theorem 2 is that it allows us to prescribe scenarios using only two initial conditions rather than three. For any scenario that starts with a small infected population, we can choose i_0 and then use $e_0 = \rho i_0$, where ρ is given by (22).

7.2   The End State

Dynamical systems often progress toward a fixed er d state. These must be states for which all of the rates of change are 0. For endemic models, there are usually only one or two such states and there are straightforward methods to determine which is the end state. The situation is much more comp icated for epidemic models. To begin, note that a fixed end state must have no further changes in R. Since $dR/dt = \gamma I$, we can only have a fixed value of R if $I = 0$. From here, we can similarly conclude that a fixed value $I = 0$ also requires $E = 0$. The chain stops here, however. There is no particular reason why a fixed end state should have $S = 0$ or any particular value of R. Based on the differential equation model (12)–(16), any state with values

$$S = s_\infty, \quad E = 0, \quad I = 0, \quad R = r_\infty = 1 - s_\infty \tag{23}$$

could serve as the end state.

In the case of the SEIR epidemic model and a few other simple ones, the end state can be found using calculus. The idea is that the relationship between the variables S and R is determined by the differential equations for those two variables. If we assume that R is a function of S, rather than being an independent function of t, we can use the chain rule to obtain an expression for dR/dS. If that expression depends only on the variable S, as is the case here, then we can use integration techniques to find the one-parameter family of anti-derivatives for dR/dS. Only one of these anti-derivatives also satisfies the initial conditions for R and S. The result is as follows:

Theorem 3 *The initial and final values of the susceptible population are related by the equation*

$$\ln s_0 - \ln s_\infty = \mathcal{R}_0(1 - r_0 - s_\infty). \tag{24}$$

Equation (24) is an analytical result that connects the initial values r_0 and s_0, the basic reproduction number $\mathcal{R}_0$, and the final value s_∞. It cannot be used to immediately calculate s_∞ for any set of input parameters because there is no way to solve the algebraic equation for s_∞. However, we can obtain several useful conclusions from the theorem:

1. The final state s_∞ depends only on the basic reproduction number and the initial conditions; it is unaffected by the values used for the rate constants η and γ. If these constants are relatively small for one scenario, then it will just take longer to reach the same final state.
2. The final state cannot have $s_\infty = 0$ because there are no values one can choose for the other parameters to satisfy (24).

3. If we want to see how s_∞ depends on the infectiousness of the disease for a given initial scenario, we can use (24) to calculate $\mathscr{R}_0$ values from given values of s_∞ and then plot a graph of s_∞ vs $\mathscr{R}_0$, without ever solving the equation for s_∞.

If we want to be able to calculate s_∞ rather than approximating it from a graph, we need a numerical method, since (24) cannot be solved for it using algebra. The equation can be manipulated in various ways to put it in the form $F(s_\infty) = 0$ for some function F. There are a variety of well-documented numerical methods for solving such equations. These methods are much harder to use in practice than they are in theory because of various complicating factors. Scientific computing software, such as MATLAB, includes built-in functions for this task. These have been written by numerical analysis experts, and you can use these if you just want an answer. If you want better mathematical understanding, you have to write your own programs to test specific methods. We include this as a possible project for the more mathematically experienced.

7.3   Section 7 Exercises

Exercise 17 (Calculus Level 0, Programming Level 0) Find estimates for the incubation period, the infectious period, and the basic reproductive number for an infectious disease of your choice. Use this information to calculate η, γ, λ, and ρ for that disease. Also estimate s_∞ assuming $r_0 = 0$ and $s_0 \approx 1$, either by using a numerical solver with (24) or by using a plot of s_∞ vs $\mathscr{R}_0$ (see consequence 3 of Theorem 3).

Exercise 18 (Calculus Level 1, Programming Level 0) As described in the opening paragraphs of Sect. 7, the graph on the left of Fig. 8 shows both the latent and infectious classes reach a peak before dropping off to 0. Note that in addition to occurring earlier in time, the latent peak is lower and approaches zero more quickly than the infectious peak. Explain why this is the case for the given parameters. Would it be possible for the latent peak to be larger than the infectious peak? Would it be possible for the latent curve to be greater than the infectious curve during certain periods of time during the epidemic? If so, for what parameters might this happen? Explain.

Exercise 19 (Calculus Level 1, Programming Level 0) Explain what the right panel of Fig. 8 suggests about the nature of the epidemiological process in the stage where the graphs of $\ln(E)$ and $\ln(I)$ are parallel. It may help to reference material in Sect. 1. Your explanation should also relate the plot to the corresponding plot in the left panel.

7.4 Section 7 Challenge Problems

Challenge Problem 3 (Calculus Level 2, Programming Level 0) Assume that $S = s_0$, $R = r_0$, $\ln I = \ln I_0 + \lambda t$, and $\ln E = \ln I + \ln \rho$ for some unknown values I_0 and ρ (note that these assumptions are justified when e_0 and i_0 are small by Fig. 8). Derive the results of Theorem 2 by substituting these assumptions into the model (12)–(15).

Challenge Problem 4 (Calculus Level 2, Programming Level 0) Assume that $\mathcal{R}_0$, η, γ, i_0, and r_0 are given, with i_0 small. Use the same method as in Challenge Problem 3 to derive a quadratic equation whose solution determines ρ. Note that this formula can be used to select an appropriate e_0 for a scenario that starts shortly after the beginning of an outbreak.

Challenge Problem 5 (Calculus Level 2, Programming Level 0) Assume R is a function of $S(t)$. Use this assumption and the model (12–15) to derive a formula for dR/dS. Then use calculus and algebra to derive the result of Theorem 3.

7.5 Section 7 Project

Research Project 6 (Calculus Level 2, Programming Level 2) Develop an algorithm for solving (24) for s_∞. If you use a method, such as Newton's method, that requires a starting guess, your algorithm should include steps that compute a good starting guess from the parameters s_0, r_0, and $\mathcal{R}_0$. Your algorithm should work for reasonable values of the parameters: for example, $1 < \mathcal{R}_0 \leq 10$.

8 Simulations

In Sect. 7, we identified five important outcomes for an epidemic model. Even when some of these can be determined analytically, there are always some that can only be determined by running a numerical simulation. We also need a numerical simulation if we want to see the full course of the epidemic rather than just seeing key outcomes.

There are two different aspects of simulation that we need to study: the numerical aspect and the modeling aspect. The distinction is best made in terms of Fig. 6. The narrow view is the realm of numerical analysis; we need to work out methods to determine the time histories of the state variables S, E, I, and R, with a given set of parameters. The broad view is the realm of modeling; we need to work out strategies for getting useful information from data for the time histories.

8.1 Numerical Simulation of Continuous Dynamical Systems

Continuous dynamical systems pose a problem for simulation because computers can only do algebra, not calculus. This means that the only way to simulate a continuous dynamical system is to approximate it by a discrete dynamical system. Consider the model $dy/dt = f(y) = by(1 - y)$, which represents an SI model, as an example. The obvious way to approximate the differential equation is to replace the derivative with the difference quotient $\Delta y/\Delta t$. Then we can define y_n to be the approximate value of y on day n. With f evaluated using the population on day n, we have

$$y_{n+1} = y_n + f(y_n)\Delta t. \tag{25}$$

(Note that $\Delta t = 1$ in our model.) This straightforward method is called *Euler's method*.[16] While this is the most intuitive way to approximate a differential equation numerically, there are much better methods. In most cases, including the models that arise in epidemiology, the method of choice is a more complicated method called the Runge–Kutta[17] method of order 4, sometimes referred to as "RK4." The basic idea of the method is that we want to retain the simple structure of Euler's method,

$$y_{n+1} = y_n + m\Delta t,$$

but with a more sophisticated estimate of the slope m. The RK4 method calculates three other slope estimates in addition to $f(y_n)$ and takes a weighted average. A more complete discussion can be found in most differential equations texts, and the formulas appear in the program *seir.m* in the Appendix.

In general, one can improve accuracy by using time intervals smaller than 1 day, but this is unnecessary when using the RK4 method with an epidemiology model because the time scales for the processes are on the order of several days.

8.2 Implementation of Numerical Simulations

We present here a brief description of the MATLAB program suite that we use to implement the numerical simulation of the SEIR model (the programs themselves appear in the Appendix). We assume that students have MATLAB installed and know how to access the software, or else know how to run MATLAB programs

---

[16] Pronounced OY-ler.
[17] Pronounced RUN-guh KUT-tah.

using the online version of Octave [12]. The reader should review the description of the individual-based model programs in Sect. 2.2.

One way to make an accessible program suite is to have it consist of a function file that does computations for one scenario and a set of driver scripts that use the computational function to do experiments. The function file *seir.m* contains the function *seir*, which requires input values for the basic reproductive number, the mean incubation time, the mean recovery time, initial conditions, and data used to implement a situation-based ending condition. The outputs of *seir* are vectors that give the values of the four population classes at each time along with the value of s_∞ calculated by applying MATLAB's built-in equation solver to solve (24).

Most of the coding in *seir.m* can be left as is when modifying the program to simulate other models. Additional function arguments and outputs may be needed, and there will likely be small changes to the *INITIALIZATION* section, but most of the changes will appear in the *FUNCTION FOR THE DIFFERENTIAL EQUATION* section. The comments in this section and the syntax of the statements in it should suffice for a novice programmer to be able to make modifications as needed.

There are three driver programs, each designed for a particular type of experiment:

- *SEIR_onesim.m* runs one simulation, leading to a plot like that in the left panel of Fig. 8. The user needs to modify the *SCENARIO DATA* section for each run.
- *SEIR_comparison.m* runs a comparison of scenarios having one parameter that takes a variety of values. There is a section for default scenario data that includes all the input parameters. Values for the parameter selected as the independent variable are specified as *xvals* in the *INDEPENDENT VARIABLE DATA* section. The first line in the computation loop assigns values from *xvals* to the desired parameter name. This uses a different value for that parameter on each run through the loop, while the other parameters keep their default values.
- *SEIR_paramstudy.m* produces plots of key outcomes as a function of one of the parameters. The parameter values are specified in almost the same manner as for *SEIR_comparison.*

8.3   Section 8 Exercises

Exercise 20 (Calculus Level 1, Programming Level 1) Our best guesses for the original strain of COVID-19 are a basic reproductive number of 5.7, an incubation period average of 5 days, and an infectious period average of 10 days. Assume an initial population that is entirely susceptible except for ten latent individuals per 100K. Run *SEIR_onesim.m* and describe what would have happened in a community that made no behavioral or public health adjustments.

Exercise 21 (Calculus Level 1, Programming Level 1) Repeat Exercise 20 using the disease you characterized in Exercise 17. Make sure that the final susceptible fraction obtained by the computer simulation matches the one you estimated in that exercise.

Exercise 22 (Calculus Level 1, Programming Level 1) The Incan Empire had a population of over one million when it was conquered by 168 Spanish Conquistadores in 1525. The Spanish had gunpowder weapons and horses, but these advantages would not have been sufficient to defeat the huge Incan army. (It took about 2 minutes to reload a single-shot arquebus.) They also benefitted by joining forces with peoples subjugated by the Incas, but that would not have happened on its own. Historians William H. McNeill and Jared Diamond have argued that the key factor in the Incan defeat was the European diseases the Spanish brought with them [4, 11]. To test this theory, set the basic reproduction number at 5, the incubation period at 12 days, and the infectious duration at 20 days, values that roughly match smallpox. Describe the effect introduction of smallpox into Incan civilization would have had, even without considering the death toll of the disease.

Exercise 23 (Calculus Level 1, Programming Level 1) For a more complete look at the effect of the basic reproduction number on epidemic progression, run *SEIR_comparison.m* using *R0* values 5, 3, 2, 1.5, and 1.25, with incubation period 5 days, infectious duration 10 days, and initial infectious fraction 0.001.

1. Discuss the graphs, explaining why the effects of *R0* are what you see.
2. Suppose a disease with *R0*=5 is combated with social distancing and measures to decrease transmission probability for each contact. What effect do you expect these social policies to have and why?

Exercise 24 (Calculus Level 1, Programming Level 1) Use *SEIR_paramstudy.m* to study the effect of the basic reproduction number on epidemic outcomes. Use *R0* values from 0 to 6 and the original default values for the other parameters. Describe and explain the results, paying particular attention to the behavior near *R0*=1.

Exercise 25 (Calculus Level 1, Programming Level 1) Use *SEIR_paramstudy.m* to do a more thorough study of the effect of the disease duration on epidemic outcomes. Use *R0*=2.5 and *tI* values from 4 to 12. Describe and explain the results. Pay particular attention to the axis limits.

8.4   Section 8 Projects

Research Project 7 (Calculus Level 1, Programming Level 1) Write an analysis of herd immunity. You can find basic information about this concept on the web, but you will need to justify all of your claims with simulation results rather than quotations from other sources. You will have to think carefully about what outcome is the most important to report. You should also address the question of why herd immunity does not work in an epidemic scenario for a disease with a large value of $\mathscr{R}_0$ (say 5.0) that starts with nearly everyone susceptible.

Research Project 8 (Calculus Level 1, Programming Level 2) How much of a problem is the assumption that transition rates are proportional to the class size? At issue is the association of this assumption with the exponential distribution. Look up the characteristics of this distribution. A more realistic choice is the Erlang E_2 distribution. An Erlang E_2-distributed process is equivalent to a sequence of two exponentially distributed stages. This means that we can get Erlang-distributed incubation times by replacing the standard SEIR model with the SEEIR model from Exercise 13. To study this new model, you will need to modify the SEIR programs for SEEIR and compare the results. Be careful about how you calculate σ; keep in mind that t_L is the average total time spent in the E classes, not the average time spent in each of the classes. Under what circumstances are we making a significant error by using the SEIR model in place of the more accurate SEEIR model?

Research Project 9 (Calculus Level 1, Programming Level 2) Research the impact of isolation of symptomatic infectious patients on the course of an epidemic. This will require you to change the model from SEIR to SEPUR, where the original infectious class has been divided into a sequence of a presymptomatic class and an unisolated symptomatic class. You can assume that some fraction q of individuals who transition out of the presymptomatic class move directly into the removed class through isolation, while the remainder move into the unisolated class and become removed in the usual way. Note that you have to be careful with your formula for the transmission

(continued)

process, since the new model has two infectious classes. Exercise 12 can serve
as a good starting point.

Research Project 10 (Calculus Level 1, Programming Level 2) Build and
study a model that incorporates social distancing and/or mask usage to
decrease transmission. Consider making the model more realistic by dividing
the population into a compliant subgroup and a noncompliant subgroup, but
keep in mind that individuals in the two groups still interact with each other.

Research Project 11 (Calculus Level 1, Programming Level 2) How does
the development of a vaccine change the course of an epidemic? You will need
a new model that accounts for several features of vaccine implementation: (1)
a new vaccine can only be distributed gradually as doses are produced and
distributed, (2) not everyone is willing to take a vaccine, and (3) not everyone
who is vaccinated develops immunity.

Acknowledgments The authors wish to thank the book editors for their assistance and encouragement, Hannah Callender Highlander for suggesting this chapter, and Julie Blackwood for her thoughtful and helpful review of the manuscript.

Appendix: Programs

hpsr.m

```
function [H,P,S,R]=hpsr(b,N,P0,S0)
%
% function [H,P,S,R]=hpsr(b,N,P0,S0)
%
%     runs a simulation of an agent-based model
%
%     H: healthy
%     P: presymptomatic
%     S: sick
%     R: recovered
%
```

```
%    b is the transmission probability
%    N is the total population
%    P0 is the initial presymptomatic population
%    S0 is the initial sick population
%
% by Glenn Ledder
% written 2020/11/29
% revised 2021/10/20
%
% direct comments to gledder@unl.edu

%% DATA

% suggested default values
% b = 1;
% N = 1000;
% P0 = 20;

%% INITIALIZATION

% limit simulation duration
maxdays = 100;

% set data structures using initial conditions
H = (N-P0-S0)*ones(1,maxdays+1);
P = P0*ones(1,maxdays+1);
S = S0*ones(1,maxdays+1);
R = zeros(1,maxdays+1);

%% COMPUTATION

for n=1:maxdays

    % calculate probability of an infected partner for a
      healthy person
    p = (P(n)+S(n))/(N-1);

    % get a 'partner' for each healthy person
    partners = rand(H(n),1);

    % count H's with infectious partners
    atrisk = length(find(partners<p));

    % get random numbers for transmission success
    exposure = rand(atrisk,1);
```

```
    % count atrisks with successful transmission
    infected = length(find(exposure<b));

    % calculate new class counts
    H(n+1) = H(n)-infected;
    P(n+1) = infected;
    S(n+1) = P(n);
    R(n+1) = R(n)+S(n);

    % check if done
    if P(n+1)+S(n+1)==0
        % delete unneeded rows
        H = H(1:(n+1));
        P = P(1:(n+1));
        S = S(1:(n+1));
        R = R(1:(n+1));
        % exit for loop
        break;
    end %if

end %for

%% END

end
```

HPSR_onesim.m

```
%% HPSR_onesim

% Plots the results of an HPSR agent-based model
  simulation

% Prints results and outcomes:
%    results is a matrix of columns for time,H,P,S,R
%    maxI is the maximum number infected
%    finalH is the ending size of H

% User specifies values for 4 parameters:
%    b is the transmission probability
%    N is the total population
%    P0 is the initial presymptomatic population
%    S0 is the original sick population
```

```
% If using Octave, add the additional option pair
   "'MarkerSize',2.5"
%   (without the double quotes) to the three plot
   statements)

% Uses hpsr.m, 2021/10/20 version

%- function [H,P,S,R]=hpsr(b,N,P0,S0)
%-
%-    runs a simulation of the HPSR agent-based disease
      model
%-
%-    H: healthy
%-    P: presymptomatic
%-    S: sick
%-    R: recovered
%-
%-    b is the transmission probability
%-    N is the total population
%-    P0 is the initial presymptomatic population
%-    S0 is the initial sick population

% by Glenn Ledder
% written 2020/11/29
% revised 2021/10/26

% direct comments to gledder@unl.edu

%% SCENARIO DATA

b = 5/6;
N = 1000;
P0 = 20;
S0 = 0;

%% INITIALIZATION

% set up for first figure

figure(1)
clf
darkgreen = [0 0.6 0];
hold on
box on
```

```
%% COMPUTATION

% collect simulation data
[H,P,S,R] = hpsr(b,N,P0,S0);

% record simulation duration [H(1) is for day 0]
days = length(H)-1;

%% OUTPUT

% get list of times for plot
times = 0:days;

% If using Octave, add the additional option pair
  "'MarkerSize',2.5"
%   (without the double quotes) to the three plot
    statements

% plot healthy, infected, recovered
plot(times,H,'d','Color',darkgreen,'MarkerFaceColor',
     darkgreen)
plot(times,P+S,'rs','MarkerFaceColor','r')
plot(times,R,'b^','MarkerFaceColor','b')
xlim([0,days])

% label axes
%    use 'FontSize',18 in Octave
xlabel('days','FontSize',14)
ylabel('populations','FontSize',14)

legend('healthy','infected','recovered','Location',
       'East')

% display results as a matrix with columns for t, H, P,
  S, R
results = [times',H',P',S',R']

% report maximum simultaneously infected
maxI = max(P+S)

% report final healthy population
finalH = H(end)

%% FIGURE 2
```

```
% uncomment the lines from 101 to get a figure of
  log(P+S)
%    add ",'MarkerSize',2.5" to the end of this plot
     command in Octave
%    use 'FontSize',18 in Octave

% figure(2)
% clf
% hold on
% box on
%
% plot(times,log(P+S),'rs','MarkerFaceColor','r')
%
% xlabel('days','FontSize',14)
% ylabel('ln(infected)','FontSize',14)
```

HPSR_avg.m

```
%% HPSR_avg

% Computes results for an HPSR agent-based model
  simulation

% Displays histograms and prints means and standard
  deviations of outcomes
%    maxI_pct is the maximum number infected by
     percentage
%    finalH_pct is the ending size of H by percentage

% User specifies values for 5 parameters:
%    b is the transmission probability
%    N is the total population
%    P0 is the initial presymptomatic population
%    S0 is the original sick population
%    numruns is the number of simulation runs in the
     experiment

% SEE SPECIAL INSTRUCTIONS (below) FOR HISTOGRAMS IF
  USING OCTAVE.

% Uses hpsr.m, 2021/10/20 version.

%- function [H,P,S,R]=hpsr(b,N,P0,S0)
```

```
%-
%-    runs a simulation of the HPSR agent-based disease
      model
%-
%-    H: healthy
%-    P: presymptomatic
%-    S: sick
%-    R: recovered
%-
%-    b is the transmission probability
%-    N is the total population
%-    P0 is the initial presymptomatic population
%-    S0 is the initial sick population

% by Glenn Ledder
% written 2020/12/30
% revised 2021/10/26

% direct comments to gledder@unl.edu

%% SCENARIO DATA

b = 1;
N = 1000;
P0 = 10;
S0 = 0;
numruns = 1000;

%% INITIALIZATION

% set up for figure

clf
for k=1:2
    subplot(1,2,k)
    hold on
end

% create row vectors for output data

maxI = zeros(1,numruns);
finalH = zeros(1,numruns);

%% COMPUTATION
```

```
for j=1:numruns
    % run simulation
    [H,P,S,~] = hpsr(b,N,P0,S0);
    % save results
    maxI(j) = max(P+S);
    finalH(j) = H(end);
end

% convert to percentages and sort
maxI_pct = 100*sort(maxI)/N;
finalH_pct = 100*sort(finalH)/N;

%% OUTPUT

subplot(1,2,1)
% For Octave, replace the histogram statement with the
   following:
% hist(maxI_pct,14,1,'FaceColor','y')
histogram(maxI_pct,'Normalization','probability',
'FaceColor','y','FaceAlpha',1/3)
xlabel('max I (pct)','FontSize',11)
ylabel('probability','FontSize',11)

subplot(1,2,2)
% For Octave, replace the histogram statement with the
   following:
% hist(finalH_pct,14,1,'FaceColor','y')
histogram(finalH_pct,'Normalization','probability',
'FaceColor','y','FaceAlpha',1/3)
xlabel('final H (pct)','FontSize',11)
ylabel('probability','FontSize',11)

% report means and standard deviations
maxI_pct_mean = mean(maxI_pct)
maxI_pct_std = std(maxI_pct)
finalH_pct_mean = mean(finalH_pct)
finalH_pct_std = std(finalH_pct)
```

seir.m

```
function [S,E,I,R,sinfty]=seir(R0,tL,tI,e0,i0,r0,target,
         maxdays)
%
% function [S,E,I,R,sinfty]=seir(R0,tL,tI,e0,i0,r0,
```

```
            target,maxdays)
%
% runs a simulation of an SEIR model
%
% S: susceptible
% E: 'exposed' ('latent' is the more accurate term)
% I: infectious
% R: removed
%
% R0 is the basic reproductive number
% tL is the mean incubation time
% tI is the mean recovery time
% e0 is the initial latent fraction
% i0 is the initial infectious fraction
% r0 is the initial immune fraction
% target is the infected fraction used as an end
  condition
% maxdays is the maximum simulation duration
%
% by Glenn Ledder
% written 2020/11/27
%
% direct comments to gledder@unl.edu

%% DATA

% suggested default values
% R0 = 5;
% tL = 2;
% tI = 10;
% e0 = 0.0001;
% i0 = 0;
% r0 = 0;
% target = 0.001;
% maxdays = 1000;

%% INITIALIZATION

% calculate parameters

eta = 1/tL;
gamma = 1/tI;
beta = R0*gamma;
s0 = 1-e0-i0-r0;
k = 1-r0-log(s0)/R0;
```

```
% set up results data structure with Y=[S,E,I,R]

results = zeros(maxdays+1,4);
Y = [s0,e0,i0,r0];
results(1,:) = Y;

y = Y';
oldx = e0+i0;

%% FUNCTION FOR Sinfty

fnc = @(s,R0,k) s-exp(R0*(s-k));  % parameterized function
fofs = @(s) fnc(s,R0,k);          % function of s alone

%% COMPUTATION

for t=1:maxdays
    % y is a column vector, Y^T
    y = rk4(1,y);
    Y = y';
    results(t+1,:) = Y;
    if Y(2)+Y(3)>min(target,oldx)
        oldx = Y(2)+Y(3);
    else
        results = results(1:(t+1),:);
        break;
    end
end

S = results(:,1);
E = results(:,2);
I = results(:,3);
R = results(:,4);

sinfty = fzero(fofs,[0,1]);

%% FUNCTION FOR rk4

    function y=rk4(dt,y0)
        % y0 is a column vector of initial conditions at t
        % y is a column vector of values at t+dt
        k1 = yprime(y0);
        k2 = yprime(y0+0.5*dt*k1);
        k3 = yprime(y0+0.5*dt*k2);
```

```
    k4 = yprime(y0+dt*k3);
    y = y0+dt*(k1+2*k2+2*k3+k4)/6;
  end
```

```
%% FUNCTION FOR THE DIFFERENTIAL EQUATION
```

```
  function yp=yprime(y)
  % split out components
      S = y(1);
      E = y(2);
      I = y(3);
  % compute derivatives
      Sp = -beta*S*I;
      Ep = -Sp-eta*E;
      Ip = eta*E-gamma*I;
      Rp = gamma*I;
  % assemble derivative
      yp = [Sp;Ep;Ip;Rp];
  end
```

```
%% END
```

```
end
```

SEIR_onesim.m

```
%% SEIR\_onesim
```

```
% Plots a comparison of population classes
```

```
% Prints results and outcomes(populations are fractions):
%    results is a matrix of columns for time,S,E,I,R,
     Delta S
%    maxI is the maximum size of class I
%    maxday is the day on which maxI occurs
%    finalS is the ending size of S
```

```
% Uses seir.m, 2020/11/27 version
```

```
% User specifies values for 6 parameters:
%    R0 is the basic reproductive number
%    tL is the mean incubation time
%    tI is the mean recovery time
%    e0 is the initial latent fraction
```

```
%    i0 is the initial infectious fraction
%    r0 is the initial immune fraction
%    target is the infected fraction used as an end
%    condition
%    maxdays is the maximum simulation duration

% Change axis label font sizes for Octave

% by Glenn Ledder
% written 2020/11/27
% revised 2021/08/01

% direct comments to gledder@unl.edu

%% SCENARIO DATA

R0 = 5;
tL = 2;
tI = 10;
e0 = .0001;
i0 = 0;
r0 = 0;

%% COMMON DATA

target = 0.001;
maxdays = 1000;

%% INITIALIZATION

clf
hold on
box on
colors = get(gca,'colororder');

%% COMPUTATION

[S,E,I,R,sinfty] = seir(R0,tL,tI,e0,i0,r0,target,
 maxdays);
days = length(I)-1;
new = S(1:days)-S(2:length(S));
new = [0;new];

%% OUTPUT
```

```
times = 0:days;

plot(times,S,'Color',colors(1,:),'LineWidth',1.4)
plot(times,E,'Color',colors(3,:),'LineWidth',1.4)
plot(times,I,'Color',colors(2,:),'LineWidth',1.4)
plot(times,R,'Color',colors(4,:),'LineWidth',1.4)

% use 'FontSize',18 in Octave
xlabel('days','FontSize',14)

% use 'FontSize',16 in Octave
ylabel('population fraction','FontSize',12)

legend('S','E','I','R','Location','West')

results = [times',S,E,I,R,new]
[M,J] = max(I);
maxI = M
maxday = J-1
finalS = sinfty
```

SEIR_comparison.m

```
%% SEIR\_comparison

% Plots a comparison of population classes for multiple
  scenarios

% Uses seir.m, 2020/11/27 version

% User specifies a list of values for one of the key
  parameters:
%    R0 is the basic reproductive number
%    tL is the mean incubation time
%    tI is the mean recovery time
%    i0 is the initial infectious fraction
%    r0 is the initial immune fraction

% e0 is calculated assuming exponential growth phase

% The program is designed so that only two lines need to
  be modified to
% make a new experiment (see '%%%' comments)
%       line 47 defines the independent variable values
```

```
%      line 79 links the independent variable name and
       values
% The legend is optional

% output figure:
%    left panel: S vs time
%    right panel: I vs time

% Change axis label font sizes for Octave

% by Glenn Ledder
% written 2020/11/27
% revised 2021/08/01

% direct comments to gledder@unl.edu

%% DEFAULT SCENARIO DATA

RC = 2.5;
tL = 5;
tI = 10;
i0 = 0.001;
r0 = 0;

%% INDEPENDENT VARIABLE DATA

%%% This section needs to be modified for each
    experiment.

%%% xvals is the set of independent variable values
xvals = [5,4,3,2];

%% COMMON DATA

target = 0.001;
maxdays = 1000;

%% INITIALIZATION

clf
for k=1:2
    subplot(1,2,k)
    hold on
    box on
    % comment out next line if desired
```

```
    axis square

    % use 'FontSize',18 in Octave
    xlabel('days','FontSize',14)
end

N = length(xvals);
finalS = zeros(1,N);
maxI = zeros(1,N);
maxday = zeros(1,N);

%% COMPUTATION and PLOTS

for n=1:N

    %%% The left side of this statement needs to be the
        independent
    %%% variable for the experiment.
    R0 = xvals(n);

    eta = 1/tL;
    gamma = 1/tI;
    beta = gamma*R0;

    a = eta;
    b = eta-gamma+gamma*R0*i0;
    c = -gamma*R0*(1-i0-r0);
    b2 = b/2;
    rho = (sqrt(b2^2-a*c)-b2)/a;
    e0 = rho*i0;

    [S,~,I,~,sinfty] = seir(R0,tL,tI,e0,i0,r0,target,
     maxdays);
    days = length(I)-1;

    [M,J] = max(I);
    maxI(n) = M;
    maxday(n) = J-1;
    finalS(n) = sinfty;

    subplot(1,2,1)
    plot(0:days,S,'LineWidth',1.4)
    subplot(1,2,2)
    plot(0:days,I,'LineWidth',1.4)
end
```

```
subplot(1,2,1)
ylim([0,1])
% use 'FontSize',16 in Octave
ylabel('susceptible','FontSize',12)

subplot(1,2,2)
% use 'FontSize',16 in Octave
ylabel('infectious','FontSize',12)
legend('R0=5','R0=4','R0=3','R0=2','Location',
'Northeast')

%% OUTPUT

maxI = maxI
maxday = maxday
finalS = finalS
```

SEIR_paramstudy.m

```
%% SEIR parameter studies

% Runs an SEIR experiment to determine how maximum
  infectious fraction and
% final at-risk fraction depend on a parameter value

% Uses seir.m, 2020/11/27 version

% User specifies a list of values for one of the key
  parameters:
%    R0 is the basic reproductive number
%    tL is the mean incubation time
%    tI is the mean recovery time
%    i0 is the initial infectious fraction
%    r0 is the initial immune fraction

% e0 is calculated assuming exponential growth phase

% The program is designed so that only a few lines need
  to be modified to
% make a new experiment (see '%%%' comments)
%    lines 49-51 define the independent variable values
%    line 54 identifies the x axis label for the graph
%    line 86 links the independent variable name and
```

```
     values

% Output figure:
%     top left panel: max infectious fraction
%     top right panel: day of max new infections
%     bottom left panel: fraction of susceptibles who
      become infected
%     bottom right panel: final fraction susceptibles

% Change x axis label font size for Octave

% by Glenn Ledder
% written 2020/11/27
% revised 2021/08/01

% direct comments to gledder@unl.edu

%% DEFAULT SCENARIO DATA

R0 = 2.5;
tL = 5;
tI = 10;
i0 = 0.001;
r0 = 0;

%% INDEPENDENT VARIABLE DATA

%%% This section needs to be modified for each
    experiment.

%%% first and last are the min and max values for the
    independent variable
%%% N is the number of points (not subdivisions)
first = .1;
last = 6;
N = 60;

%%% xname is the name for the x axis label
xname = 'R0';

%% COMMON DATA

target = 0.001;
maxdays = 1000;
```

```
%% INITIALIZATION

clf
for k=1:4
    subplot(2,2,k)
    hold on
    box on

    % use 'FontSize',18 in Octave
    xlabel(xname,'FontSize',14)
end
colors = get(gca,'colororder');

% xvals holds whatever values are being used for the
  independent variable
xvals = linspace(first,last,N);

finalS = zeros(1,N);
fracI = zeros(1,N);
maxI = zeros(1,N);
maxday = zeros(1,N);

%% COMPUTATION

for n=1:N

    %%% The left side of this statement needs to be the
        independent
    %%% variable for the experiment.
    R0 = xvals(n);

    eta = 1/tL;
    gamma = 1/tI;
    beta = gamma*R0;

    a = eta;
    b = eta-gamma+gamma*R0*i0;
    c = -gamma*R0*(1-i0-r0);
    b2 = b/2;
    rho = (sqrt(b2^2-a*c)-b2)/a;
    e0 = rho*i0;
    s0 = 1-e0-i0-r0;

    [~,~,I,~,sinfty] = seir(R0,tL,tI,e0,i0,r0,target,
     maxdays);
```

```
    [M,J] = max(I);
    maxI(n) = M;
    maxday(n) = J-1;
    finalS(n) = sinfty;
    fracI(n) = (s0-sinfty)/s0;
end

%% OUTPUT

subplot(2,2,1)
plot(xvals,maxI,'Color',colors(3,:),'LineWidth',1.7)
ylabel('max fraction infectious')
subplot(2,2,2)
plot(xvals,maxday,'k','LineWidth',1.7)
ylabel('days for max infectious')
subplot(2,2,3)
plot(xvals,fracI,'Color',colors(2,:),'LineWidth',1.7)
ylabel('fraction infected')
subplot(2,2,4)
plot(xvals,finalS,'Color',colors(1,:),'LineWidth',1.7)
ylabel('final fraction susceptible')
```

References

1. Blackwood, J.C., Childs, L.M.: An introduction to compartmental modeling for the budding infectious disease modeler. *Letters in Biomathematics* (2018) https://doi.org/10.1080/23737867.2018.1509026
2. Brauer, F.: Mathematical epidemiology is not an oxymoron. *BMC Public Health* (2009) https://doi.org/10.1186/1471-2458-9-S1-S2
3. Davis, C.P.: *Swine Flu*
   https://www.medicinenet.com/swine_flu/article.htm. Cited 11 December 2020.
4. Diamond, J.: *Guns, Germs, and Steel*. W.W. Norton, New York (1997)
5. Gammack, D., Schaefer, E., Gaff, H.: Global dynamics emerging from local interactions: agent-based modeling for the life sciences, in *Mathematical Concepts and Methods in Modern Biology: Using Modern Discrete Models*, ed Robeva, R. & Hodge, T. Academic Press (2013)
6. Homp, M. and Ledder, G.: *The Mathematics of an Illness Outbreak: part 1, Disease Simulation Activity* (2019)
   https://drive.google.com/drive/folders/1k7ImUJEuO3hz5uSZ1hPVwv0KSm9aSEjs. Cited 29 November 2020.
7. Jungck, J.R., Gaff, H., Weisstein, A.E.: Mathematical manipulative models: In defense of "Beanbag Biology". *CBE-Life Sciences Education*, **9**, 201–211 (2010)
8. Kermack, W.O., McKendrick, A.G.: A contribution to the mathematical theory of epidemics. *Proc Royal Soc London*, **115**, 700–721 (1927) https://doi.org/10.1098/rspa.1927.0118.
9. Khan, S.: *Derivative as a Concept* (2017)
   https://www.youtube.com/watch?v=N2PpRnFqnqY. Cited 29 November 2020.
10. Ledder, G.: *Mathematical Modeling for Epidemiology and Ecology*, 2nd ed. Springer-Verlag, New York (2022)

11. McNeill, W.H.: *Plagues and Peoples*. Anchor Press, Garden City, NY (1976)
12. *Octave Online*
    https://octave-online.net/. Cited 11 December 2020.
13. Railsback, S.F., Grimm, V.: *Agent-based and Individual-based Modeling: A Practical Intro-duction*, 2nd edition. Princeton University Press, Princeton (2019)
14. Sanche, S., Lin, Y.T., Xu, C., Romero-Severson, E., Hengartner, N., Ke, R.: High contagious-ness and rapid spread of severe respiratory syndrome coronavirus 2. *Emerging Infectious Diseases*, **26**: 1470–1477 (2020) https://doi.org/10.3201/eid2607.200282.
15. Smith, G.: *Back of the Envelope Modelling of Infectious Disease Transmission Dynamics for Veterinary Students*. Cambridge Scholars Publishing, Newcastle-Upon-Tyne (2019).
16. Wikipedia: *Lotka-Volterra Equations*
    https://en.wikipedia.org/wiki/Lotka-Volterra_equations. Cited 29 November 2020.
17. Wikipedia: *Moore's law*
    https://en.wikipedia.org/wiki/Moore's_law. Cited 29 November 2020.

Playing with Knots

Allison Henrich

Abstract

One of the best ways to get acquainted with a mathematical object is to play with it. Knots are no exception. In the process of trying to learn more about the properties of knots, a team of undergraduate researchers played around with knot diagrams, changing crossings to see how the knottedness of their knots was affected. It felt like they were playing a game. And so they were! It turns out that the team invented a game that is now called the "Knotting-Unknotting Game," one of the first topological-combinatorial games of its kind to be studied. Since this initial game was introduced, many more researchers have devised games that can be played with knot and link diagrams. Studying these games to determine their mathematical properties can be just as fun as simply playing them, and in the process, we can learn more about the knotty objects we're playing with.

Suggested Prerequisites *An introductory proofs course would be a helpful prerequisite for this research, although several students have done research on knot games without formal training in proof techniques. Otherwise, you can learn everything you need to know along the way.*

1 Knots: Knotted and Unknotted

When did you first learn that mathematicians study knots? Last year? Last week? Today? *What?! Mathematicians study knots??!!*

A. Henrich (✉)
Seattle University, Seattle, WA, USA
e-mail: henricha@seattleu.edu

It's rather surprising since we typically think of knots as being very practical (think: shoelaces, sailing, camping). Sometimes, we focus on their beauty, as we do with Celtic knots. But rarely do we think of knots as being mathematical. It turns out, however, that knot theory is an important area of study in mathematics.

1.1   What Is a Knot? (The Basics)

What exactly *is* a knot from a mathematician's perspective? The answer is similar to what your intuition tells you, with one possible exception. A **knot** is a nice, smooth *closed curve* sitting in space in such a way that it doesn't intersect itself, although it might wind around itself in any number of interesting and intricate ways. How your intuition might diverge from the mathematical notion is just this: we don't allow knots to have loose ends. If we did, any interesting knotting behavior happening in our string might simply slip off one of the ends. More generally, a **link** is a—possibly intertwining—collection of knots, where each knot in the collection is called a **component** of the link. See Fig. 1 for an example of a knot and a 2-component link.

Knots are often represented by knot diagrams. Formally, a **knot diagram** is a 2-dimensional closed curve with a finite number of self-intersections where one strand of the knot crosses another. Crossings in knot diagrams are decorated so that it is clear which strand should pass over and which should pass under at each intersection point. The special knot that can be represented by a diagram with no crossings is called the **unknot**. All other knots can be referred to as **nontrivial knots**.

Similarly, a **link diagram** is a 2-dimensional representation of a link. If a link can be represented by a diagram where there are no crossings occurring between different components of the link, then we say that the link is **splittable**. If, on the other hand, a link's components are inextricably intertwined (i.e., not splittable), we call it **unsplittable**. The simplest splittable links are simply disjoint collections of unknots, called **unlinks**. In Fig. 2, we illustrate the unknot and give an example of a nontrivial knot called the **trefoil** and a nontrivial (unsplittable) link called the **Hopf link**.

Fig. 1 A knot (left) and a link (right) created by Rob Scharein using KnotPlot (www.knotplot.com)

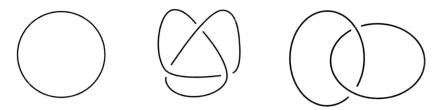

Fig. 2 A diagram of the unknot (left), the trefoil knot (center) and a Hopf link (right)

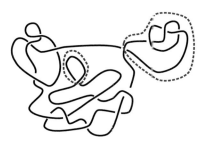

Fig. 3 An alternating knot with four nugatory crossings

There are a number of questions we can ask related to knots, links, and their diagrams. For instance, how can knot and link diagrams be used to provide a measure of complexity for knots and links? One possible answer can be found in the notion of a crossing number. The **crossing number** of a knot or link is the minimum number of crossings needed to represent it diagrammatically. We view knots and links with lower crossing numbers as being simpler and those with higher crossing numbers as being more complex. Of course, any knot or link—no matter how simple—can be represented by a diagram that has an enormous number of crossings, so we often describe knots and links in terms of their simplest diagrams. A diagram of a knot or link that actually *achieves* the minimum crossing number is called a **minimal crossing diagram** of the knot or link. The unknot is the unique knot with crossing number 0. In fact, it is the only knot with crossing number less than 3!

Minimal crossing diagrams all have one thing in common. They don't have any nugatory crossings. A **nugatory crossing** is a crossing such that it is possible to draw a (wobbly) circle through the crossing that splits the knot diagram into two parts: a part that is entirely outside of the circle and a part that is entirely inside the circle. In other words, this (wobbly) circle only touches the knot diagram at that one crossing. In Fig. 3, we see a knot diagram with four nugatory crossings. Two are encircled. Can you find the other two?

Here's a thought experiment! Try to visualize how this diagram could be simplified by twisting portions of the knot that are connected by nugatory crossings in order to remove these unnecessary crossings. Do you think the resulting diagram is a minimal crossing diagram? If not, why not? If so, can you prove it?

While minimal crossing diagrams of knots certainly exist, it can be quite difficult to *prove* that a given knot diagram is a minimal crossing diagram. Thus, it can

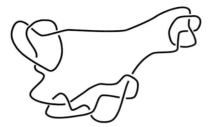

Fig. 4 An alternating knot with no nugatory crossings, derived from the diagram in Fig. 3 by untwisting the knot at its nugatory crossings. This knot has crossing number 11

be hard to compute a knot's crossing number and, hence, determine the knot's complexity. Fortunately, there is a special kind of knot diagram for which we have an easy test for minimality: an alternating knot diagram. A knot diagram is **alternating** if, as you travel around the diagram, you travel over, then under, then over, then under, etc. at each of the crossings. Any knot that has an alternating diagram is called an **alternating knot**. The diagram in Fig. 3 is alternating, and so represents an alternating knot. However, we've already noted that this diagram is not a minimal crossing diagram because of those pesky nugatory crossings. But if we were able to untwist the knot to remove the nugatory crossings from our diagram, we'd be in good shape according to a famous result: the Tait Conjecture.[1]

Theorem 1 (Tait Conjecture, Thistlethwaite-Kauffman-Murasugi) *Any alternating knot diagram that has no nugatory crossings is a minimal crossing diagram of the knot it represents.*

The Tait conjecture tells, for instance, us that the "untwisted" version of our knot from Fig. 3, shown in Fig. 4, is a minimal crossing diagram.

Now, about this "untwisting." We need a formal way of describing how we can manipulate knot diagrams without fundamentally altering our knot. It's about time we learned about knot equivalence!

1.2 Knot Equivalence and Reidemeister Moves

The central question of knot theory deals with classification: when are two knots or links the same and when are they different? What does it mean for two knots to be "the same," or, equivalent? We say two knots (or links) are **equivalent** if one can be deformed into the other in space simply by stretching, bending, pulling,

---

[1] While it was once a conjecture, this Tait Conjecture is now a theorem! In fact it is one of many theorems proven from conjectures made by Peter Guthrie Tait. We still call Tait's conjectures "conjectures" for historical reasons.

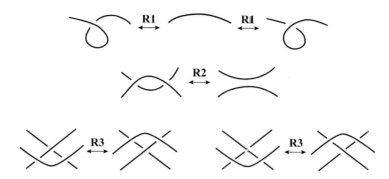

Fig. 5 The Reidemeister moves: R1, R2, and R3. To visualize how these moves work, imagine that the curve involved in one of the moves is a piece of a larger knot or link. The move alters just this piece, leaving everything else unchanged

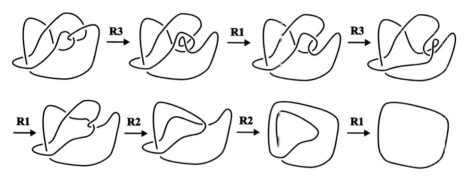

Fig. 6 Simplifying a non-minimal unknot diagram using R-moves

expanding, contracting, etc. *as long as* the knot never passes through itself or gets cut and reattached at any point.

One tool knot theorists use to help them determine if two knots or links are equivalent is Reidemeister's theorem. In the 1920s, Kurt Reidemeister [18] and, independently, Alexander and Briggs [4] proved that two knot or link diagrams represent the same knot or link *if and only if* their diagrams can be related by a sequence of Reidemeister moves (shown in Fig. 5) and planar isotopies (i.e., "wiggling"). Notice that the R1 move adds or removes a little kink in the knot or link. The R2 move overlaps two strands or pulls two strands—where one lies entirely above the other—apart. The R3 move passes a strand of the knot over or under a crossing. Figure 6 shows how Reidemeister moves can be used to simplify a complicated unknot diagram.

If you think about it, Reidemeister's theorem is a surprising result. It makes sense that these three moves don't change the knot or link type, but what is quite unexpected is that these moves completely describe knot equivalence—no other moves are needed to pass from one knot or link to another equivalent knot or link.

Fig. 7 An unknot in disguise

Fig. 8 Several members of the twist knot family: T_1, T_2, T_3, and T_4

As we'll see later, the three Reidemeister moves are particularly useful when we need to identify whether or not a given knot or link diagram represents an unknot or unlink.

Exercise 1 Use the Reidemeister moves to show that the knot diagram in Fig. 7 is an unknot.

Challenge Problem 1 Show that any knot diagram with one or two crossings is unknotted.

1.3 Some Useful Knots and Links

Now that we've used Reidemeister moves to better acquaint ourselves with the unknot, let's get to know a few other knots. One knot family that has some interesting properties is called the **twist knot** family. Several members of the family are pictured in Fig. 8. Notice that the first twist knot, T_1, is a trefoil. The second twist knot in the list, T_2, also has a special name: it's called the **figure-eight knot**. All twist knots are made up of a clasp (with two crossings) and a twist (that can have as many crossings as we like!). Figure 9 shows how this works. As we will soon see, twist knots are particularly nice to play with. This is because if you were to change one of the crossings in the clasp of any twist knot by making the overstrand the understrand—no matter how many crossings are in your twist knot diagram overall—you'd get a complicated diagram of the unknot.

Exercise 2 Prove the following. For any twist knot, if one of the clasp crossings in its standard diagram is changed so that the understrand instead passes *over* the crossing, the resulting knot is the unknot.

clasp crossings

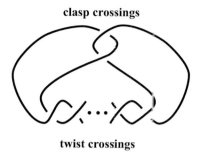

twist crossings

Fig. 9 A standard diagram for a general twist knot

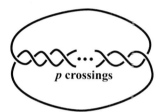

p crossings

Fig. 10 A general $(2, p)$-torus knot

Fig. 11 The $(2, 5)$-torus knot. Image courtesy of Mitch Riehling

Another knot type we like to play with is the $(2\ p)$-torus knot, shown in Fig. 10. This knot is called a $(2, p)$-torus knot because we could view it as a string wrapping around a torus (think: doughnut) two times in one direction and p times in another without touching itself. See Fig. 11 for an illustration of this phenomenon for the $(2, 7)$-torus knot. Note that p is always an odd number for the $(2, p)$-torus knots. If p is even, we have a $(2, p)$-torus *link*.

Whereas the twist knots are very easy to unknot if we are allowed to change one of the crossings in the clasp, the $(2, p)$-torus knots are very difficult to unknot by changing crossings. The number of crossings that need to be changed to unknot a $(2, p)$-torus knot grows as p grows.

Both the twist knots and the $(2, p)$-torus knots (and links) are special cases of a more general type of knot and link called a **rational knot** or **rational link**. Rational knots and links are formed by taking two strings and twisting them together, alternating horizontal and vertical twists, and then gluing together the loose ends

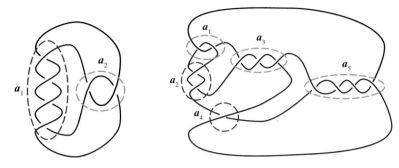

Fig. 12 The rational knots $(a_1, a_2) = (4, 2)$ [left] and $(a_1, a_2, a_3, a_4, a_5) = (-2, 2, 3, 1, -4)$ [right]

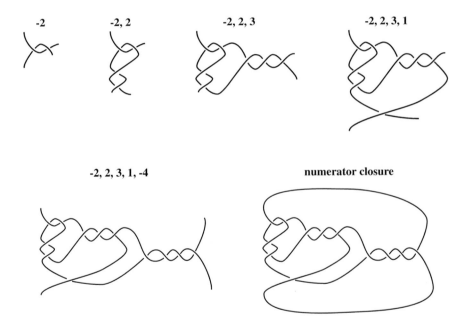

Fig. 13 Building rational knot $(a_1, a_2, a_3, a_4, a_5) = (-2, 2, 3, 1, -4)$

in a particular way (using what is called the "numerator closure"). In Fig. 12, we see two examples of rational knots—one formed by an even number of twists and one formed using an odd number of twists.

Figure 13 shows how the second example in Fig. 12 is formed by twisting and gluing. In particular, this example shows how we can form the rational knot associated with the 5-tuple $(a_1, a_2, a_3, a_4, a_5) = (-2, 2, 3, 1, -4)$ by taking a horizontal negative 2-twist (a_1), then adding a vertical positive 2-twist below it (a_2), followed by a horizontal positive 3-twist to the right (a_3), a vertical positive 1-twist below it (a_4), and then a horizontal negative 4-twist to the right (a_5). Here, the sign of a twist is determined by whether the overstrand of each of the crossings has a positive or negative slope; a positive slope is associated with a positive twist.

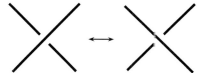

Fig. 14 A crossing change

The example from Fig. 13 demonstrates how to form a rational knot or link using an odd number of twists. If our rational knot or link is formed from an even number of twists, as in the first example shown in Fig. 12, we follow the same process, except we begin with a vertical twist rather than a horizontal twist.

Notice that twist knots are all the rational knots of the form $(a_1, 2)$ (*turn your head sideways!*) and $(2, p)$-torus knots and links are all the rational knots and links of the form (a_1).

1.4   Unknotting Operations and Numbers

Knots in closed loops can get quite complicated with no hope of simplification. That is, there is no hope *unless* we are prepared to change the knot by cutting and pasting it at various points. Such moves that allow us to simplify the knotting in a knot are called **unknotting operations**. The most fundamental example of an unknotting operation is one we've already been playing with: a **crossing change**. (See Exercise 2.) A crossing change can be performed on a knot diagram, and it usually changes the *topological type* of the knot (i.e., transforms it into a non-equivalent knot). We see this crossing change operation pictured in Fig. 14.

An interesting fact about knots that a beginning knot theorist might prove early on in their investigation is the following: Any knot can be unknotted with crossing changes.

Challenge Problem 2   Provide an argument to show that any knot diagram can be transformed into a diagram of the unknot by performing some number of crossing changes.

Once we know that crossing changes can unknot *any* knot, we might ask a follow up question: *how many* crossing changes do you need? The unknotting number answers this question. For a given knot K, the **unknotting number** of K is the minimum over all possible diagrams, D, of K of the number of crossing changes that need to be performed in D in order to produce a (possibly quite complicated) diagram of the unknot.

You've shown without too much work that the unknotting number of any twist knot is 1. But determining the unknotting number of a knot is not always so easy. What makes the unknotting number a difficult number to determine is that sometimes a more complicated diagram of a knot requires fewer crossing changes to unknot it than a simpler one. The pair of diagrams in Fig. 15 was discovered independently by Nakanishi [17] and Bleiler [5]. Both Nakanishi and Bleiler verified that the diagram on the left cannot possibly be unknotted with fewer than three crossing changes. But the diagram on the right, which is related to the first diagram by an R2 move that adds two crossings, only requires two crossing changes to unknot. This knot is called the $(5,1,4)$-pretzel knot, $P_{5,1,4}$, because it can be formed by connecting a vertical 5-crossing twist, a 1-crossing twist, and a vertical 4-crossing twist by joining the twist ends at the tops and bottoms.

Exercise 3 Verify that (i) the minimal crossing diagram of $P_{5,1,4}$ shown in Fig. 15 on the left can be unknotted by changing its three highlighted crossings and (ii) the diagram on the right in Fig. 15 can be unknotted by changing its two highlighted crossings.

To help describe which strand goes over and which strand goes under in a knot or link diagram more easily, we should introduce a couple of new concepts: (1) an orientation of a knot or link and (2) the sign of a crossing. First of all, we say that a knot or link is **oriented** if it's like a one way street, i.e., it has a preferred direction in which we should travel around it. We can denote this in our diagram using an arrow. Next, using an orientation for our knot or link, we can define the **sign of a crossing**. Imagine looking at a crossing in such a way that the overstrand is pointed upwards from our perspective. If the understrand is traveling from right to left, the crossing is **positive**. If the understrand is traveling from left to right, the crossing is **negative**. See Fig. 16.

Suppose a knot is given to us without an orientation, and we randomly choose an orientation for it. It turns out that all of the crossings will have the same signs

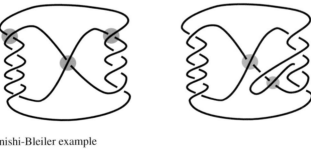

Fig. 15 Nakanishi-Bleiler example

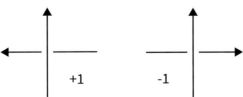

Fig. 16 The sign of a crossing

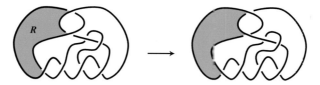

Fig. 17 The region crossing change operation applied to the shaded region, R

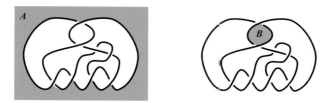

Fig. 18 Regions A and B

as if we had chosen the other possible orientation. So, crossings signs are well-defined even for unoriented knots. The same is not exactly true for links. If we have an oriented link and we change the orientation of one or more (but not all) of the components, it will likely change some crossing signs. Before reading on, convince yourself of these facts!

The crossing change operation—i.e., the operation that changes the sign of a crossing in a knot or link diagram—will become very important for several of the games we describe below, but another important unknotting operation is called the **region crossing change** (RCC) move. The fact that this operation can transform any knot diagram into a diagram of the unknot was only recently discovered by Ayaka Shimizu [19]. Here's how it works. To perform a region crossing change on a knot diagram, we first select a region R. Then, we perform the crossing change operation on all crossings *along the boundary* of R. An example is pictured in Fig. 17.

Exercise 4 What would the diagram in Fig. 17 look like if we had instead performed an RCC move on region A (the exterior region) or B, pictured in Fig. 18?

While we might be able to convince ourselves that individual crossing changes can unknot knots, it is really quite surprising that region crossing changes can do the same since we can only change groups of crossings, not individual crossings. An enlightening proof of this fact can be found in [19].

Exercise 5 Find a collection of regions for the diagram on the left of Fig. 17 such that, if you perform RCC moves on this set of regions, you get an unknot diagram. (Note that the order in which you perform two RCC moves doesn't affect the resulting diagram.)

Just as with the ordinary crossing change operation, we can define an unknotting number for the region crossing change. We have to define it a bit differently, though, for it to be an interesting quantity to study. It was proven that every knot has a (perhaps very complicated) diagram such that only one region needs to be changed to unknot it [3]. So, we define the **region unknotting number** as the minimum number of RCC moves required in any *minimal crossing* diagram to unknot the knot. See [19] to learn more.

Exercise 6 Prove that the region unknotting number of any twist knot is 1.

Fun fact! If we think about these unknotting operations for links, something curious happens. Any link can be transformed into the unlink with some number of crossing changes, as you might expect, but there are some links that can't be unlinked using the region crossing change. The Hopf link, shown in Fig. 2, is an example of such a link.

Exercise 7 Convince yourself that the Hopf link diagram shown in Fig. 2 *cannot* be transformed into the unlink by performing RCC moves.

1.5   Knot Invariants

When you're given two knot or link diagrams, often you'd like to know if they are equivalent or not. If you can find a sequence of Reidemeister moves that gets you from one to the other, that's great! They're equivalent! But what if you can't? They might be equivalent—perhaps you just haven't figured out the right sequence of moves. But if they're distinct, how can you *prove* it? For instance, how can we prove that the unknot and the trefoil are actually non-equivalent knots?[2] The answer is simple: Use knot invariants!

A **knot invariant** is a function that takes as its input a knot—or, more likely, a knot diagram—and outputs a value. This value might be "yes" or "no." It might be a natural number or an integer. It could be a polynomial. It might even be a far more complex algebraic object like a *graded abelian group*. (If you want to know what those are, you might be interested in math grad school!) The point is that when you input two knots into a knot invariant, if the two outputs are different, that gives you proof that the two inputs were different knots. If the two outputs are the same, however, you might need to compute a different invariant to see if you get a different result.

We've actually already seen three different knot invariants: the crossing number, the unknotting number, and the region unknotting number of a knot. These invariants all have natural numbers as output values—so in that sense they are nice—but they

---

[2] Actually, you can prove they're distinct using the Tait conjecture we learned about in Sect. 1.1, but this conjecture is proven using the Kauffman bracket, a relative of the famous knot and link invariant called the Jones polynomial.

are tricky to compute. Let's say you can unknot a knot diagram with two crossing changes. Does that mean the unknotting number of that knot is 2? Not necessarily. It could be 1 or even 0 (if it's the unknot in disguise). Ruling out these options can be tricky, requiring theorems that use other knot invariants to provide lower bounds on the unknotting number.

For the games described in this chapter, it is often important to prove that something is unknotted or knotted, linked or unlinked. Sometimes, invariants like the Jones polynomial or the linking number can be helpful. You can learn about these in [1, 11], if needed. However, there are two wonderful tools that we can use to do our work for us, identifying when knots are knotted using the Jones polynomial and other invariants. Both are web-based and involve drawing a knot freehand. The first tool, called the Knot Identification Tool, was created by Joshua Horowitz and can be found here: http://joshuahhh.com/projects/kit/. This program can identify knots with up to 10 crossings. The second, called KnotFolio, was created by Kyle Miller and can be found here: https://kmill.github.io/knotfolio/. Why not navigate to one of these sites now and explore before moving on?

Congratulations! You are now a beginner knot theorist! Now that you have a basic foundation in knot theory, we can turn our attention to the main event: *games on knots and links*!

2   The Knotting-Unknotting Game

Most of the games involving knots and links that we'll be studying in the rest of this chapter can trace their origins back to the Knotting-Unknotting Game. It's curious, then, that the Knotting-Unknotting Game was invented by accident! A group of undergraduate researchers—Noel MacNaughton, Sneha Narayan, Oliver Pechenik, and Jen Townsend—were doing knot theory research one summer at the Williams College SMALL Research Experiences for Undergraduates (REU) program. One day, they were trying to determine how much over/under crossing information you can forget in a diagram of an unknot and still know that your diagram is unknotted. While trying to solve their problem, it felt to them like they were playing a game. And so they were! So, they formalized the rules of the game and studied game strategy for a few knot families as a side project for the rest of the summer. In addition to the paper they wrote on their primary research [15], they wrote a second paper [14] introducing the Knotting-Unknotting Game and many others. Later, another undergraduate, Will Johnson, expanded on their results in [16].

The Knotting-Unknotting Game can be played as follows. Two players are given a **knot shadow**, that is, a knot diagram where all of the over/under crossing information is missing. (Aside: When a crossing's under/over information is missing, we call it a **precrossing**. Also, if a knot diagram has some crossings and some precrossings, we call it a **pseudodiagram** rather than a diagram or a shadow. In Fig. 19, we see several examples of pseudodiagrams.) One player, the Unknotter (U), wants to turn this shadow into a diagram of the unknot, while the other player, the Knotter (K), wants to make the shadow into a diagram of

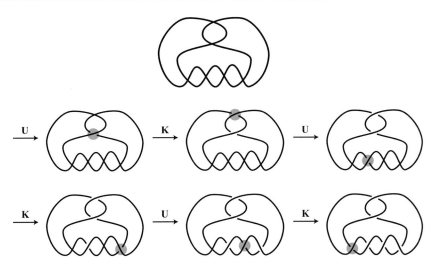

Fig. 19 The Knotting-Unknotting Game is played on the T_4 twist knot. The Unknotter, U, plays first. The final diagram is a complicated diagram of a figure-eight knot. Since this is a nontrivial knot, K has won the game

a nontrivial knot. During game play, U and K take turns choosing precrossings. When a precrossing is chosen by a player, the player can decide which strand will go over and which will go under at the crossing. In other words, the player chooses **crossing information**. This process of determining crossing information for a precrossing is called **resolving** a precrossing. When a player's turn is over, the crossing information they have just chosen is now fixed. The game continues until all precrossings have been resolved, and the knot type of the final diagram determines the game's winner.

We give one example of game play in Fig. 19 to give the reader an idea of how the game works. In this example, the Unknotter plays first on a twist knot with four twist crossings. She starts in the clasp. The Knotter follows up in the clasp, making the crossings so that they're alternating to avoid an R2 move that would effectively hand the Unknotter a win. The rest of game play takes place in the twist precrossings. In the end, the twist simplifies, but doesn't fall apart altogether. The resulting knot is the figure-eight knot, which is nontrivial. So, the Knotter wins! To see other examples of this game being played, we recommend reading [8] and [14].

There are a few questions we should be asking at this point.

1. K won. Is that because U played poorly, or did K have a *winning strategy*?
2. What would have happened if K had gone first? Could U have won?

Both of these questions are ultimately asking about winning strategies, so let's take a moment to talk about outcomes for these types of games. All of the games we're going to be studying in this chapter are combinatorial games. This means that there is one player who can always guarantee themselves a win if they play according to a winning strategy. A **winning strategy** in a game is an optimal strategy that guarantees that a given player will win the game, regardless of how their opponent plays. (Note that one or more moves that follow a winning strategy may seem counterintuitive, given the player's goal!)

Now, for the twist knot shadow shown in Fig. 19, did K have a winning strategy or did U simply fail to play optimally? It turns out that K did indeed have a winning strategy, so it wasn't the Unknotter's fault that she lost. Let's see how to prove it.

Since K is the player with a winning strategy and K plays second, we first need to consider all possible opening moves that U might play and decide how K should respond to each one. We can simplify the possibilities we need to consider by observing that playing in one of the clasp precrossings is equivalent to playing in the other, and playing in any of the twist precrossings is equivalent to playing in any other twist precrossing. So, we need only decide what to do if U plays in the clasp or the twist.

Let's say we're playing as K, the Knotter. If U plays in the clasp, it is critical that we also play in the clasp. Otherwise, U can play again in the clasp and ensure that an R2 move can be performed to unclasp the clasp and unknot the entire knot. So, we must play in the clasp in such a way that the crossings are alternating, avoiding an R2 move.

Otherwise, if U plays in the twist, we should resolve a precrossing in the twist so that it has the same sign as U's crossing. Why is this? Notice that, no matter how you orient the strands involved in an R2 move, one of the crossings is positive and one is negative. If our goal is to keep the twist as twisted up as possible, we want to avoid the situation that there are two positive and two negative crossings in the twist (which would allow for two R2 moves to completely untwist it). So, we'd like to make sure that at least three of the four crossings have the same sign.

In fact, these options for how we respond to U's first move also tell us how to play the remainder of the game. We should play to ensure that the clasp is clasped and the twist doesn't get too untwisted (that is, we need to ensure at least three crossings have the same sign).

Suppose we play according to the strategy outlined above. What are the outcomes? Well, if the clasp is alternating and the twist has four positive or four negative crossings, the knot is T_4 (if the entire diagram is alternating) or T_3 (if not). If the clasp is alternating and the twist has three positive or three negative crossings that means a single R2 move is possible. The resulting knot is the figure-eight knot, T_2, (if the entire diagram is alternating) or the trefoil, T_1, (if not). In any of these cases, the outcome is a nontrivial knot, so K wins.

Exercise 8 Suppose that the two crossings in the clasp of the twist knot shadow shown in Fig. 19 are resolved to be positive as are three of the four twist crossings. Use Reidemeister moves to show that the resulting knot is the trefoil by showing it can be reduced to an alternating, 3-crossing diagram.

Exercise 9 Invite a friend who has not been told the above winning strategy to play the Knotting-Unknotting Game with you on the twist knot from Fig. 19. Ask your friend to go first and play the role of the Unknotter. As the Knotter playing second, test out the winning strategy outlined above to see how it works.

Now we know what happens when we play the Knotting-Unknotting Game on a shadow of a standard T_4 twist knot diagram, assuming the Knotter plays second. But what if the Knotter plays first? In fact, the Unknotter has a winning strategy in that case. (*Before reading any further,* can you prove it?)

More generally, is there something special about T_4 or do similar strategies work for all twist knots? Yes and no. The precise answer is contained in the following result.

Theorem 2 ([14]) *Suppose the Knotting-Unknotting Game is played on a standard shadow of a twist knot, T_n, with $n + 2$ crossings. Then, the Knotter has a winning strategy if and only if n is even and the Knotter plays second.*

To prove this theorem, we make the following observations.
Observations for the Proof of Theorem 2:

1. If the Unknotter can play second in the clasp precrossings, the Unknotter can win.
2. The Unknotter can play second in the clasp precrossings if
   a. n is even and the Unknotter plays second, or
   b. n is odd and the Unknotter plays first.
3. One player can respond to the other player's moves in the twist precrossings so that R2 moves can remove all but possibly one (n odd) or two (n even) crossings in the twist, so the $n = 1, 2$ strategies determine the remaining cases.

Challenge Problem 3 Prove Theorem 2 by proving the observations above and finding winning strategies in the base cases of T_1 and T_2.

In [14], there are examples of knot families where the Unknotter wins both playing first and second. There are also examples where the second player always wins, regardless of whether the second player is U or K. Similarly, there are examples where the first player always wins, regardless of whether the first player is U or K. But there are no examples where the Knotter wins both playing first and second.

Research Project 1 Find an infinite family of knot shadows for which the Knotter has a winning strategy playing the Knotting-Unknotting Game— regardless of whether he plays first or second—or prove that none exists.

Research Project 2 Familiarize yourself with the knot shadows for which we know winning strategies from [14, 16], and [7]. Find a knot shadow or family of knot shadows that has not been studied and determine who has a winning strategy when playing the Knotting-Unknotting Game.

3   The Region Unknotting Game

When we described how to play the Knotting-Unknotting Game, we said the game board should be a knot shadow. There is an equivalent (albeit less intuitive) way we could have described the game. The game can be played on a knot diagram rather than a shadow. In this case, a player's move involves choosing a crossing and deciding whether to keep it as is (i.e., "fix it") or perform a crossing change. Once a crossing has been chosen (whether or not it has been changed), it is removed from game play. This is a little more cumbersome way to play since you have to keep track of which crossings the players have visited. But it does have one advantage. We can see how to invent a new, related game using a different unknotting operation!

In [6], undergraduates Sarah Brown, Frank Cabrera, Riley Evans, Gianni Gibbs, graduate student James Kreinbihl and myself introduced a variation on the Knotting-Unknotting Game that uses *region* crossing changes instead of ordinary crossing changes. In our game, there are still two players, U and K, with the same unknotting and knotting goals. The players still take turns making moves on a knot diagram. But instead of performing crossing changes or fixing crossings, the players can perform region crossing changes (aka RCC moves) or "fix" regions in the plane by removing them from game play. In other words, on a player's turn, they pick a region for which they either change all the crossings or fix all the crossings along the boundary of that region. Once this choice has been made, that region cannot be chosen again. This new game is called the **Region Unknotting Game**.

We give a sample game in Fig. 20. In this example, U moves first, performing an RCC move on a twist region. This immediately unknots the diagram, but it doesn't stay unknotted for long. On his first move, K responds by changing a region joining the twist and the clasp. The Unknotter then fixes another twist region. Game play continues until all of the regions have been chosen. In the end, four (blue) regions were changed using RCC moves and four (purple) regions were fixed. The resulting knot is nontrivial, and so K wins.

Once again, we'd like to know if the fact that K won in the sample game was a fluke or if K actually has a winning strategy on this knot diagram. The following theorem provides an answer. Since the Knotter played second and played well, he was destined to win!

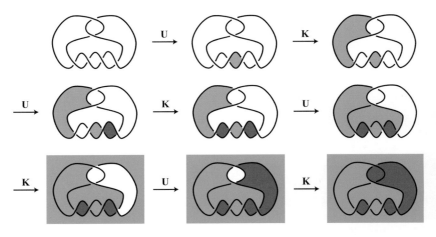

Fig. 20 The Region Unknotting Game is played on the T_4 twist knot. Regions highlighted in blue are regions on which an RCC move has been performed. Those regions in purple are regions that have been chosen, but fixed. In this game, the Unknotter, U, plays first. The final diagram is a complicated diagram of a trefoil, so K has won the game

Theorem 3 ([6]) *If the Region Unknotting Game is played on a standard alternating diagram of twist knot, T_n, where n is even, then the second player has the winning strategy, regardless of which player plays second.*

We prove one half of this theorem in the following lemma.

Lemma 1 *Suppose the Region Unknotting Game is played on a standard alternating diagram of twist knot, T_n, as in Fig. 21. If n is even and the Unknotter plays second, then the Unknotter has a winning strategy.*

Proof The Unknotter, U, playing second, wins in this game by ensuring that the clasp can be unclasped using an R2 move. She achieves this by ensuring that crossings a and b have the opposite signs. Notice that the only region that has a but not b along its boundary is region 5, and the only region that has crossing b but not a along its boundary is region 4. Changing any other region does not affect whether or not crossings a and b have the same sign. The Unknotter's strategy is to ensure that a and b have opposite signs, so moves on regions other than on 4 or 5 are not relevant for game play. Since the Knotter, K, plays first and the knot diagram has an even number of regions, U will play last. Thus, U can force K to play first on the pair of regions 4 and 5. This ensures that U can play on the corresponding region in such a way that crossings a and b have opposite signs at the end of game play. The resulting knot is the unknot. Hence, the Unknotter wins.

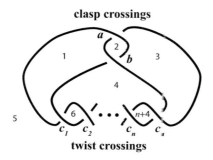

Fig. 21 A labeled twist knot

Theorem 3 takes care of the case when our twist knot diagram has an even number of crossings, but what happens if the number of crossings is odd? We have the following result.

Theorem 4 ([6]) *If the Region Unknotting Game is played on a standard alternating diagram of twist knot, T_n, where n is odd and the Unknotter plays first, then the Unknotter has a winning strategy.*

Challenge Problem 4 Prove Theorem 4.

Now, we've dealt with the twist knot cases where n is even and K goes first, n is even and U goes first, and n is odd and U plays first. Which case are we missing? It turns out that the authors of [6] weren't able to figure out what happens in the missing case. Perhaps you can!

Research Project 3 Find which player has a winning strategy if the Region Unknotting Game is played on a standard alternating diagram of twist knot, T_n, where n is odd and the Knotter plays first.

In [6], the authors also studied some specific diagrams of $(2, p)$-torus knots. In particular, they proved several results for diagrams of the $(2, 3)$, $(2, 5)$, and $(2, 7)$-torus knots. They were unable to prove a more general result, however. This brings us to our next open problem.

Research Project 4 Find which player has a winning strategy if the Region Unknotting Game is played on the standard $(2, p)$-torus knot diagram pictured in Fig. 10 for all odd p.

Finally, there are many more knot diagrams out in the world than the twist and torus knot diagrams we've been looking at. Explore the universe of knot diagrams to tackle the following problem!

Research Project 5 Find an infinite family of knot diagrams for which you can determine who has a winning strategy in the Region Unknotting Game.

4 The Linking-Unlinking Game

The Knotting-Unknotting Game not only led to the invention of the Region Unknotting Game, but it also inspired Professor Adam Giambrone and undergraduate Jake Murphy to invent yet another game: the Linking-Unlinking Game [10]. In this game, players are given a link shadow rather than a knot shadow. Just as before, they take turns resolving precrossings. In this game, however, one player—the Unlinker (U)—wants to create a *splittable* link (i.e., a link where the components are not linked with each other) while the other player—the Linker (L)—aims to create an *unsplittable* link. (In a variant of this game, one may modify the desired outcomes so that the Unlinker must produce the *unlink*.)

Let's take a look at a sample game. In Fig. 22, we see the Linking-Unlinking Game being played on a shadow of the rational link given by the triple $(2, 2, 4)$. The Unlinker moves second and wins. Her strategy is to move adjacent to where the Linker moves and resolve crossings so that an R2 move can simplify the diagram after each of her moves. This is a fantastic winning strategy for the Unlinker that actually extends to an infinite family of rational links.

Challenge Problem 5 In the sample game shown in Fig. 22, the Unlinker wins moving second. Show that if the Linker moves second on this link shadow, the Linker has a winning strategy.

Theorem 5 ([10]) *Suppose we have a shadow of a rational 2-component link defined by the n-tuple* $(a_1, ..., a_n)$.

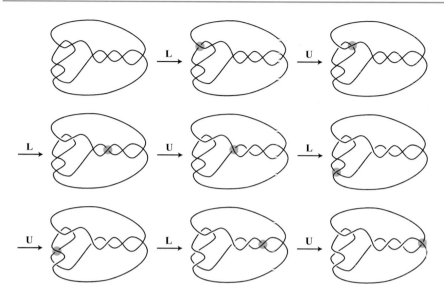

Fig. 22 The Linking-Unlinking Game is played on a rational link

1. *If $a_{2k+1} = 0$ for all k, then the Unlinker has a winning strategy.*
2. *If either*
   a. $a_{2k+1} \neq 0$ for at least one k and all of the a_i are even,
   b. $n = 2$ and both a_1 and a_2 are odd, or
   c. $n \geq 3$, both a_1 and a_n are odd, and all other a_i are even,
   then the second player has a winning strategy.

As we can see, the sample game falls into the category of 2(a), so the second player has a winning strategy, which agrees with what we saw in the example and Challenge Problem 5. Note that Theorem 5 refers to the possibility that numbers in the n-tuple defining a rational link might be 0. This corresponds to simply replacing a twist in the directions for creating a rational knot or link diagram (as in Sect. 1.3) by two parallel, nonintersecting strands. In Fig. 23, we see an example of this phenomenon that also satisfies the first condition in Theorem 5. This link, given by the 5-tuple $(0, 2, 0, 1, 0)$, is a modification of the diagram of rational knot $(-2, 2, 3, 1, -4)$ shown in Fig. 12 where we've replaced all odd twists by 0 twists. As we can see, this link starts out as an unlink, so there is no possibility for an L win.

Giambrone and Murphy proved additional theorems about rational links, but they also showed the following more general result.

Theorem 6 ([10]) *Suppose S is the shadow of a 2-component link.*

1. *If the only precrossings in S are self-intersections within each component (i.e., the diagram is already split), then the Unknotter wins.*

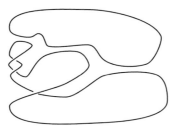

Fig. 23 The rational link associated with the 5-tuple $(0, 2, 0, 1, 0)$

2. *If S contains (i) at least one precrossing that involves both components and (ii)
   an odd number of precrossings that are self-intersections of either component,
   then the Linker has a winning strategy when playing first.*
3. *If the shadow contains (i) at least one precrossing that involves both components
   and (ii) an even number of precrossings that are self-intersections of either
   component, then the Linker has a winning strategy when playing second.*

Notice that the result you proved in Challenge Problem 5 is an example of Case 3,
since the shadow you were given has (six) precrossings involving both components
and two precrossings that are self-intersections of one of the components.

You may have spotted that there are some missing cases in Theorem 6. These
missing cases provide us with open problems.

Research Project 6 Suppose you are playing the Linking-Unlinking Game
and the Linker plays second. Find a family of 2-component link shadows for
which you can determine a winning strategy such that each shadow contains
(i) a nonzero number of precrossings that involve both components and (ii) an
odd number of precrossings that are self-intersections of either component.

Research Project 7 Suppose you are playing the Linking-Unlinking Game
and the Linker plays first. Find a family of 2-component link shadows for
which you can determine a winning strategy such that each shadow contains
(i) a nonzero number of precrossings that involve both components and (ii) an
even number of precrossings that are self-intersections of either component.

5   The KnotLink Game

Next up: the KnotLink Game! This game combines a variant of the Linking-Unlinking Game with the Knotting-Unknotting Game to create a game that can be played on either a knot shadow or a link shadow. The KnotLink Game was introduced in 2020 by undergraduate Hunter Adams, high school student Solden Stoll, and yours truly in [2].

In this game, the Simplifier, S, plays with the goal of creating either an unknot or an unlink, and the Complicator, C, aims to make any nontrivial knot or link. Most moves in the game consist of players taking turns resolving precrossings, but each player has a special move they have the option to use *at most once* at any point in the game. This special move is called **smoothing** a precrossing and is pictured in Fig. 24. A player may opt to smooth a precrossing either vertically or horizontally, as shown in the figure. Notice that smoothing a crossing can transform a knot into a link and vice versa, which is why our players need to prepare to play on both types of pseudodiagrams.

We provide a sample game in Fig. 25. The Simplifier and the Complicator face off on our favorite type of knot shadow, the shadow of a minimal crossing twist knot diagram. Here, the Complicator moves first. The Complicator begins by resolving

Fig. 24 Smoothing a precrossing. We assume that outside of this local picture, the rest of the knot or link diagram is unchanged

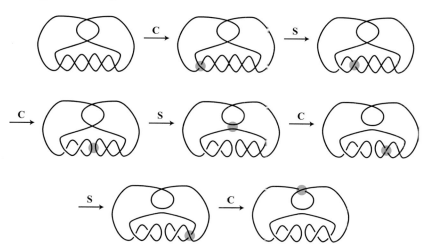

Fig. 25 A sample KnotLink game

a crossing in the twist. The Simplifier responds by resolving an adjacent crossing in the twist in such a way that a simplifying R2 move can be performed. Next, the Complicator moves on another crossing in the twist, this time performing a smoothing. The Simplifier then uses their one smoothing move in the clasp. In the remaining moves, players take turns resolving the remaining crossings. Finally, a diagram of the unknot is produced and the Simplifier wins.

Could the Complicator have won this game if he had played more strategically? Not if the Simplifier was playing according to her winning strategy! Recall that twist knots are the special types of rational knots associated with the ordered pair $(a_1, 2)$. Given this observation, the following theorem tells us that the Simplifier should be the winner on the twist knot shadow in Fig. 25.

Theorem 7 ([2]) *If the KnotLink Game is played on a standard (a_1, a_2) rational knot diagram where a_1 is odd and a_2 is even, then the Simplifier wins, regardless of whether they move first or second.*

The previous theorem has implications for odd twist knots, but what about even twist knots? Even twist knots are the rational knots given by ordered pairs of the form $(a_1, 2)$ where a_1 is even. The following theorem tells us that the situation is a bit different in the even twist knot case.

Theorem 8 ([2]) *If the KnotLink Game is played on a standard (a_1, a_2) rational knot shadow where both a_1 and a_2 are even, then the second player has winning strategy regardless of their goal.*

Let's look at how to prove half of Theorem 8.

Lemma 2 ([2]) *If the KnotLink Game is played on a standard (a_1, a_2) rational knot shadow where both a_1 and a_2 are even and the Complicator moves first, then the Simplifier has winning strategy.*

Proof First, we observe that *which* crossing a player moves on within a twist does not affect game strategy or game outcomes, so without loss of generality, we move on the first crossing in each twist. We proceed by cases, considering all of the possibilities for how the Complicator might play on his first move. Diagrams corresponding to each case are shown in Fig. 26.

Case 1: Suppose the Complicator performs a horizontal smoothing in the a_2 twist. Then, the Simplifier should perform a vertical smoothing in the a_2 twist. Now, neither player has the smoothing move available to them, and there are an even number of crossings in the a_1 twist and the a_2 twist remaining. For the rest of the game, when the Complicator moves in a_2, the Simplifier should move in a_2, resolving any crossing in any way since the a_2 twist crossings no longer affect the linking of the diagram. Moreover, when the Complicator resolves a crossing in a_1, the Simplifier should respond by resolving an adjacent crossing in a_1 so

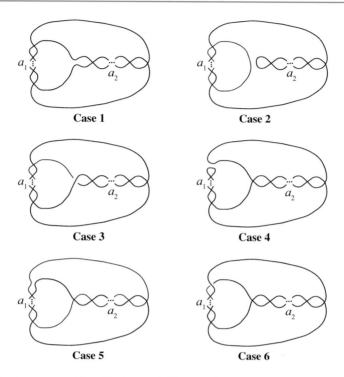

Fig. 26 The six cases considered in the proof of Lemma 2

that an R2 move can be performed to remove the two crossings from the diagram. This will result in an unlink diagram.

Case 2:   Suppose the Complicator performs a vertical smoothing in the a_2 twist. Then, the Simplifier should perform a horizontal smoothing in the a_2 twist and then follow the same strategy as in Case 1.

Case 3:   Suppose the Complicator resolves a crossing in the a_2 twist. Then the Simplifier should resolve an adjacent crossing in a_2 so that an R2 move can be performed, reducing the game to a game on a smaller, $(a_1, a_2 - 2)$ rational knot (if $a_2 > 2$) or an unknotted pseudodiagram (if $a_2 = 2$).

Case 4:   Suppose the Complicator performs a horizontal smoothing in the a_1 twist. This case is analogous to Case 2 and should be handled similarly.

Case 5:   Suppose the Complicator performs a vertical smoothing in the a_1 twist. This case is analogous to Case 1 and should be handled similarly.

Case 6:   Finally, suppose the Complicator resolves a crossing in the a_1 twist. This case is analogous to Case 3 and should be handled similarly.

Challenge Problem 6   Prove the other half of Theorem 8. In other words, prove the following: If the KnotLink Game is played on a standard (a_1, a_2) rational knot shadow where both a_1 and a_2 are even and the Simplifier moves first, then the Complicator has winning strategy.

Theorems 7 and 8 have the following corollary.

Corollary 1 *Suppose the KnotLink Game is played on a standard diagram of a twist knot, T_n, with $n + 2$ crossings. Then the Complicator has a winning strategy if and only if n is even and the Simplifier moves first.*

In [2], winning strategies are also determined for the case of a rational knot given by (a_1, a_2) where both a_1 and a_2 are odd as well as for $(2, p)$-torus knots and links. There are many ways that these results might be extended. For instance, as of this writing, nobody has yet done the following.

Research Project 8 Determine which player has a winning strategy in the KnotLink Game played on a rational knot or link given by the triple (a_1, a_2, a_3) for each combination of parities (i.e., even/odd properties) for the a_i values.

Research Project 9 Determine winning strategies in the KnotLink Game played on more complex rational knots, i.e., rational knots given by n-tuples $(a_1, a_2, \ldots, a_n)$ where $n > 3$.

Research Project 10 Determine winning strategies in the KnotLink Game played on shadows of your favorite non-rational knots.

6   The Link Smoothing Game

We now go all-in on using smoothings as moves in knot and link games! The Link Smoothing Game, introduced by Inga Johnson and myself in [12], is played on a connected knot or link shadow. Two players take turns selecting precrossings in the shadow, as usual, but in the Link Smoothing Game, players *smooth* at precrossings instead of resolving them. There are two possible choices for how to smooth at a precrossing (just as in Fig. 24), and these choices can have quite different effects on the diagram. The goal of one player, the Knotter (K), is to keep the diagram connected so that the final game board is a simple unknot while the goal of the other player, the Linker (L), is to disconnect the diagram so that the final game board is an unlink with two or more components.

A sample game where the Knotter plays first is shown in Fig. 27. Notice that after the Knotter's initial move, every time the Linker makes a move, he is attempting to set up a situation where a portion of the diagram can be disconnected from the rest of the diagram. (On each of his moves, he creates an R1-like loop.) So, the Knotter is forced to play defense for the rest of the game, although she is able to effectively thwart the Linker's attempts to disconnect the diagram. This is a common feature of Link Smoothing Games in which the Knotter wins.

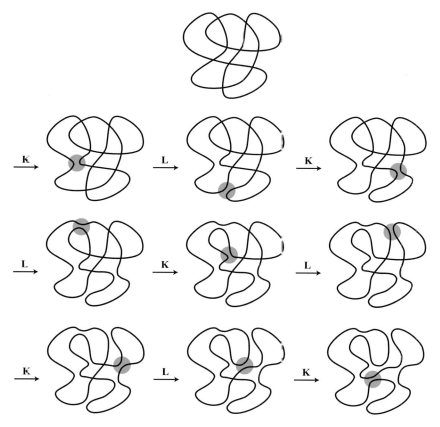

Fig. 27 A sample Link Smoothing game where the Knotter wins by producing a final diagram that has a single component

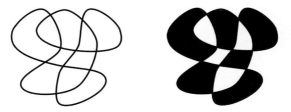

Fig. 28 Example of a shadow and its checkerboard coloring

Fig. 29 The "black" graph for the example in Fig. 28 is shown in red (left), and the "white" graph is shown in blue (center). The black-and-white graphs are shown together (right) to demonstrate their duality

Exercise 10 Find a friend and play the Link Smoothing Game on the standard shadow of any twist knot, T_n, with any $n \geq 3$. Which player has a winning strategy on your twist knot shadow? Explain why.

The Link Smoothing game was studied in [12] and again in [13] not by playing games on the link shadows themselves, but instead by analyzing game play on the *checkerboard graph* associated with a link shadow. Knot theorists have for years used the fact that the shadow of a knot or link can be represented using a planar graph to get a handle on a knot's underlying structure [1]. This representation provides tremendous insight into the game strategy for the Link Smoothing Game.

Sounds good! So, *what is a checkerboard graph again?*

The **checkerboard graph**, G, of a knot or link shadow D is constructed by first black-and-white checkerboard coloring the regions of D, as in Fig. 28. Then, the "black" graph is formed by placing a vertex in each black region. Each crossing in D determines an edge that connects the vertices of the crossing's two bordering black regions. The "white" graph is constructed similarly, but with vertices coming from white regions, including the outside region. See Fig. 29. The black-and-white graphs of a shadow may contain loops or multiple edges between a given vertex pair.

We should note that these two graphs are *duals* of each other. This means that the white graph can be determined from the black graph or vice versa. We simply place vertices in the regions of the black graph's diagram and place one edge perpendicular to each edge in the black graph so that it joins the vertices in the regions it connects. The graph we've just created is the white graph! See Fig. 29.

Fig. 30 The link shadow after the first move of the game in Fig. 27 (left) together with its black graph (center) and white graph (right)

Since we're interested in using these graphs to study the Link Smoothing Game, we need to understand how the moves in the game affect the black-and-white graphs. If a smoothing is used that joins two black regions (as in Fig. 30), then the two vertices in those regions merge into one, contracting along the edge that connects them through the smoothed crossing. Notice that a smoothing that joins two black regions breaks a connection between two white regions, so an edge in the white graph gets deleted. If, on the other hand, a smoothing that connects two white regions is used during game play, the same phenomena will occur. This time, however, two white graph vertices will contract into one, and one black graph edge will be deleted.

When we play the Link Smoothing Game on the checkerboard graphs, the Knotter is the winner if she can keep both graphs connected. The Linker wins if either graph becomes disconnected at any point during game play. Now, we can play the Link Smoothing Game on the black-and-white graphs rather than on the link shadow!

Exercise 11 Show how the black-and-white checkerboard graphs are transformed with each move in the sample game from Fig. 27.

Exercise 12 Following your observations from Exercise 10, see how game play on the twist knot shadow proceeds when played on the corresponding checkerboard graphs. What feature of the twist knot's black graph do you suspect gives one player the upper hand in this game?

Using the link-graph correspondence, the following theorem was proven in [12, 13].

Theorem 9 ([12, 13]) *Let G be a checkerboard graph associated with a connected link shadow D. Graph G represents a game board on which the second player has a winning strategy when playing the Link Smoothing Game if and only if G is composed of two edge-disjoint spanning trees.*

Fig. 31 A graph (left) with two of its cycles (dashed edges; center, right)

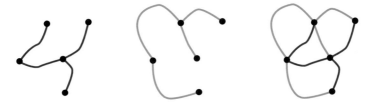

Fig. 32 Two trees (left, center); a graph decomposed into two edge-disjoint spanning trees (right)

We see that this is a powerful theorem, but what in the world does it mean for a graph to be *composed of two edge-disjoint spanning trees*? Good question! (We didn't even know the answer when we began studying the Link Smoothing Game!) Let's take a moment to gather a few definitions from graph theory.

First of all, a **cycle** in a graph is a path *(think: going on a walk in a graph, making sure you never visit the same place twice)* that starts and ends at the same vertex. Two examples of cycles are shown in Fig. 31. A **tree** is a connected graph with no cycles. A **spanning tree** of a graph G with n vertices is a subset of $n - 1$ edges of G that form a tree. In Fig. 32, we see two trees, and we see how they're both spanning trees for the graph from Fig. 31. In fact these trees are **edge-disjoint spanning trees** because they don't share any edges.

We see in Fig. 32 that the game board we get after playing the first move in our sample game (Fig. 27) fits the criteria for Theorem 9. So the second player has a winning strategy. If we begin our game one move in, the Knotter plays second, so the Knotter does indeed have a winning strategy. What about the original game board from the sample game? Well, the Knotter is the *first* player in this game, so the Knotter can ensure that the first move of the game puts her into a situation where Theorem 9 guarantees her a win. Thus, The Knotter has a winning strategy when playing first on the original knot shadow as well!

For more on the Link Smoothing Game, we encourage you to read [12].

7   Conclusion

Many of the major questions about the Link Smoothing Game have already been answered, but we included it here to inspire you to invent new, and perhaps similar games. For instance, another interesting smoothing game called the Region

Smoothing Swap Game was inspired by this work. It is a hybrid of the Link Smoothing Game and the Region Unknotting Game. We refer you to [13] to learn more about this more complex game that can not only be played on a flat surface, but also on tori, tori with more than one hole, and even nonorientable surfaces like Klein bottles!

In fact, there are even more games you can play on knot diagrams that aren't included here. One of these is a two-player game, where players' moves involve performing Reidemeister moves [9]. Another is a single-player game—more of a puzzle—called Region Select [20]. This game is based on the results about the region crossing change operation from [19].

These are all interesting games to learn more about and explore. For now, though, we leave you with your biggest challenge and the most fun open problem in this chapter. Enjoy playing!

> **Research Project 11** Taking inspiration from all the games you learned about here, invent your own game that can be played on knot and link diagrams. Does your game have two players? Or perhaps your game only has one player... or three! Be creative. Don't edit your ideas for possible new games before you've explored many possibilities. Once you have settled on a game concept, explore your game. Which game boards are interesting to study? If your game is a two-player game, which players have winning strategies when moving first or second on which game boards?

Acknowledgments The author would also like to thank the Simons Foundation (#426566) for their support of this work. She would also like to thank Steve Klee and the anonymous reviewer for their valuable suggestions as well as the editors of this volume.

References

1. C. Adams, *The Knot Book: An Elementary Introduction to the Mathematical Theory of Knots*, American Mathematical Society, Providence, RI, 2004.
2. H. Adams, A. Henrich and S. Stoll, The KnotLink Game. *FUMP J. Undergrad. Res.* **3** (2020) 110–124.
3. H. Aida, Unknotting operations of polygonal type. *Tokyo J. Math.* **15** (1992) 111–121.
4. J. W. Alexander, G. B. Briggs, On types of knotted curves, *Ann. of Math.*, **2** no. 28 (1926), 562–586.
5. S. A. Bleiler, A note on unknotting number, *Math. Proc. Camb. Phil. Soc.*, **96** no. 3 (1984), 469–471.
6. S. Brown, F. Cabrera, R. Evans, G. Gibbs, A. Henrich and J Kreinbihl, The region unknotting game. *Math. Mag.* **90** no. 5 (2017) 323–337.

7. T. Chang, On the Existence of a Winning Strategy in the T(4, 3) Knotter Vs. Unknotter Game. (2013) https://www.semanticscholar.org/paper/On-the-Existence-of-a-Winning-Strategy-in-the-T%284%2C-Chang/fbd80141d764c3be26dd1623dd518ddbf0de6501?p2df.
8. M. Cohen and A. Henrich, A knot game with not KNerds, *Math Horiz.* **20**:2 (2012) 25–29.
9. S. Ganzell, A. Meadows, and J. Ross, Twist Untangle and Related Knot Games. *Integers* **14** (2014) G4.
10. A. Giambrone and J. Murphy, The linking-unlinking game. *Involve.* **12**, no. 7 (2019) 1109–1141.
11. I. Johnson and A. Henrich, *An Interactive Introduction to Knot Theory.* Courier Dover Publishers (2017).
12. A. Henrich and I. Johnson, The link smoothing game. *AKCE Int. J. Graphs Comb.* **9** no. 2 (2012) 145–159.
13. A. Henrich, I. Johnson, and J. Ostroff, The region smoothing swap game. *Osaka J. Math.* **58** no. 1 (2020).
14. A. Henrich, N. MacNaughton, S. Narayan, O. Pechenik, R. Silversmith, and J. Townsend, A midsummer knot's dream, *College Math. J.* **42** no. 2 (2011) 126–134.
15. A. Henrich, N. MacNaughton, S. Narayan, O. Pechenik, and J. Townsend, Classical and virtual pseudodiagram theory and new bounds on unknotting numbers and genus, *J. Knot Theor. Ramif.* **20** no. 04 (2011) 625–650.
16. W. Johnson, Combinatorial game theory, well-tempered scoring games, and a knot game. *arXiv:1107.5092 [math.CO].* (2011) 1–279.
17. Y. Nakanishi, *Unknotting numbers and knot diagrams with the minimum crossings*, Mathematics Seminar Notes, Kobe University, **11**, (1983), 257–258.
18. K. Reidemeister, Elementare Begründung der Knotentheorie, (German) *Abh. Math. Sem. Univ. Hamburg* **5** no. 1 (1927), 24–32.
19. A. Shimizu, Region crossing change is an unknotting operation, *J. Math. Soc. Japan* **66**:3 (2014) 693–708.
20. A. Shimizu, A game based on knot theory, *Asia Pacific Mathematics Newsletter* **2** (2012) 22–23.
21. I. Torisu, On nugatory crossings for knots, *Topology Appl.* **92**:2 (1999) 119–129.